THERMAL ENERGY STORAGE ANALYSES AND DESIGNS

THERMAL ENERGY STORAGE ANALYSES AND DESIGNS

PEI-WEN LI
Department of Aerospace and Mechanical Engineering,
The University of Arizona (USA)

CHO LIK CHAN
Department of Aerospace and Mechanical Engineering,
The University of Arizona (USA)

Academic Press is an imprint of Elsevier
125 London Wall, London EC2Y 5AS, United Kingdom
525 B Street, Suite 1800, San Diego, CA 92101-4495, United States
50 Hampshire Street, 5th Floor, Cambridge, MA 02139, United States
The Boulevard, Langford Lane, Kidlington, Oxford OX5 1GB, United Kingdom

Library of Congress Cataloging-in-Publication Data
A catalog record for this book is available from the Library of Congress

British Library Cataloguing-in-Publication Data
A catalogue record for this book is available from the British Library

ISBN: 978-0-12-805344-7

For information on all Academic Press publications
visit our website at https://www.elsevier.com/books-and-journals

Publisher: Joe Hayton
Acquisition Editor: Lisa Reading
Editorial Project Manager Intern: Gabriela D. Capille
Editorial Project Manager: Mariana Kuhl
Production Project Manager: Sruthi Satheesh
Cover Designer: Matthew Limbert

Typeset by SPi Global, India

CONTENTS

PREFACE

Since most types of renewable energy are highly intermittent and need to be stored to serve the need of heat or electrical power in a timely manner, thermal storage is becoming a key area of research and development due to its capability of storing an unlimited amount of energy, particularly for concentrated solar thermal power generation to provide an extended period of operation after sunset.

Having been involved in experimental test and mathematical analysis for thermal energy storage for concentrated solar thermal power systems for a long time, the authors feel an obligation to summarize their work and put it into book form, to help engineers in the renewable energy industry to better understand the physics of thermal energy storage and to address various engineering issues. To accomplish this objective, *Thermal Energy Storage Analysis and Design* considers the significance of thermal energy storage systems over other systems designed to handle large quantities of energy storage. The book introduces the different thermal storage system structures and the configuration of thermal storage material and heat transfer fluids. It provides surveyed properties for heat transfer fluids and thermal storage materials (for sensible or latent heat storage), and presents mathematical modeling and analysis of the heat transfer for thermal charging and discharging processes. Computer code is provided with the mathematical analysis implemented as a tool. Approaches and steps for thermal storage system design and sizing (either hot or cold storage) using mathematical tools for heat transfer analysis are presented in the book for engineering applications.

The authors gratefully acknowledge the diligent work of several graduate students, Jon Van Lew, Wafaa Karaki, M.M. Valmiki, Ben Xu, and Eric Tumilowicz, who have contributed to the better understanding of the physics of thermal energy storage and have also provided the full implementation of the mathematical analysis modeling into computer code.

Pei-Wen Li, Cho Lik Chan
University of Arizona, Tucson, AZ, United States

CHAPTER 1

Introduction

Contents

Abstract

Energy storage is critical to the development of renewable energy technologies in the future, due to the fact that almost every type of renewable energy is irregular and intermittent regarding its availability and magnitude. Among various energy storage approaches, thermal storage is one of the most promising large-scale energy storage technologies. Here in the introduction the list of topics and technical issues for thermal energy storage are briefly presented as a reader's overview.

Keywords: Energy storage technology, Thermal energy storage, Large-scale energy storage, Concentrated solar power, Industrial energy technology

1.1 THE SIGNIFICANCE OF ENERGY STORAGE TO RENEWABLE ENERGY TECHNOLOGIES

Fossil fuels have been our major energy resources in the past, driving the industrialization and the modernization of human society. There is no doubt that the world economy will continue to largely rely on fossil fuels, such as coal, oil, natural gas, and atomic energy as well, far into the future. More importantly, the demand for energy in general will continue to rise [1]. However, it also has been widely recognized that it is impossible for the worldwide production of coal and oil to keep rising with no limitation. The world has to find new energy resources, or otherwise the current energy technologies have to be improved dramatically so that energy consumption is reduced significantly. The world also has recognized that the heavy use of fossil fuels causes pollution of the air and environment, which could cause hardship and suffering for future generations.

Thermal Energy Storage Analyses and Designs
http://dx.doi.org/10.1016/B978-0-12-805344-7.00001-8

The need for clean energy resources and environmental protection drives the strong demand for developing renewable energy, particularly solar energy and wind energy, around the world. Renewable energy must represent a significant proportion of our energy package in the future, in order for the world economy and environment to have sustainable development.

Almost all types of renewable energy, particularly solar energy and wind power, do not have very high energy density and are available only intermittently or even irregularly. However, the utilization of energy and power in industry, homes, and the workplace often has a different phase compared to the time/period of availability of renewable energy. Therefore, to provide renewable energy and power and follow the load or demand, energy storage and smart energy dispatch technologies are critically needed. Energy storage mechanisms in different forms at a variety of levels of capacities, or quantities, and different periods of time must be better understood. The degradation or loss of stored energy as a function of time will largely determine the energy storage efficiency and length of time. Technologies capable of long-term energy storage and low degradation are very challenging but are in great demand.

1.2 A BRIEF INTRODUCTION TO ENERGY STORAGE TECHNOLOGIES

The following energy storage technologies have been widely recognized and studied by scientists and engineers in the energy engineering field:

(1) Thermal energy storage including both hot storage and cold storage.
(2) Direct storage of electricity in capacitors or batteries.
(3) Electrical and mechanical energy storage using flywheels.
(4) Pumped hydroelectric energy storage relying on reservoirs.
(5) Compressed air energy storage.
(6) Electrical energy storage using a combination of electrolyzers and hydrogen fuel cells.

The availability and success of these energy storage technologies will significantly affect the success of the renewable energy industry and the directions of technology development. Taking solar energy as an example, the use of solar energy has three major objectives: to produce electricity, make fuels, and directly collect and use heat, as shown in Fig. 1.1. To make electricity from solar energy, photovoltaic panels and solar thermal power plants are often used, which is strategy I in Fig. 1.1. For short-term energy storage,

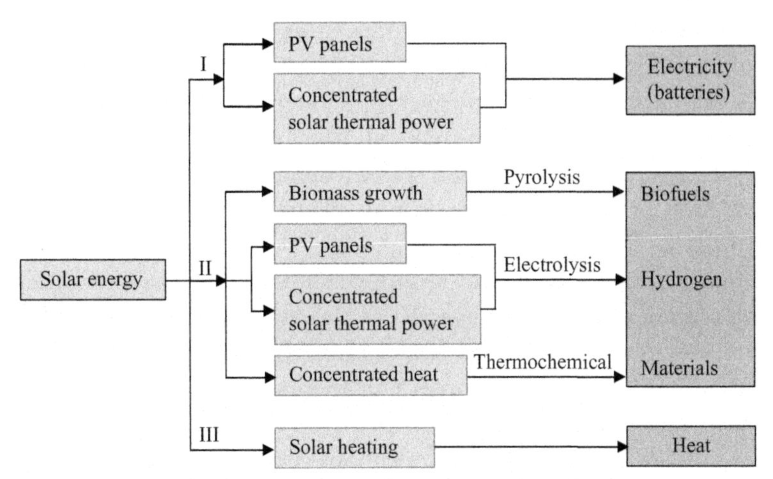

Fig. 1.1 Solar energy for three purposes of energy supply and storage [1].

an auxiliary energy storage system can store electricity directly or store thermal energy in a solar thermal power plant. However, for longer-term energy storage or for fuel-energy storage, one needs to consider strategy II in Fig. 1.1. Biofuels, hydrogen, or fuel materials can be stored very long term, and thus using renewable energy for fuel and hydrogen production is receiving more attention recently. Direct use of solar energy for heating can also be assisted with thermal storage, which is categorized as strategy III in Fig. 1.1.

Of all the energy storage technologies, direct electrical energy storage is the most desirable; for example, the electricity generated by solar PV panels or wind turbines requires storage for dispatch. However, energy storage based on direct electricity storage using batteries and capacitors is limited to small to medium capacity levels, such as supplying energy for portable electronic devices, computers, electrical motorcycles and cars for use in limited operations and over a short time. Electrical and mechanical energy storage using flywheels can also provide a small to medium amount of energy capacity. Storing a very large quantity of energy, such as in a power plant to meet the needs of several hundred megawatts of electrical power output for 4–8 h, is only possible using thermal energy storage, pumped hydroelectric energy storage, or compressed air energy storage. Because of the need for elevated water reservoirs, pumped hydroelectric energy storage is only available in a small number of limited locations. Storing a sufficiently large quantity of energy using compressed air energy storage requires a huge volume for high pressure air, which is only realistic if large-volume natural caverns are available. The technology of electrical energy storage using a combination of electrolyzers and hydrogen fuel cells is considered to a long-term energy storage technology. However, it requires hydrogen storage, and this technology still needs more research and development. Therefore, thermal energy storage turns to be the most viable technology that is readily available with no limitation on location or quantity of energy storage for a short time, up to 12 h.

Thermal energy storage can be either hot or cold storage, for different periods of time and with various scales of capacity. Hot thermal energy storage can provide heating, while cold thermal energy storage may provide cooling as needed at another time. Hot thermal energy storage is widely applied in industry, such as for concentrated solar thermal power generation, thermal storage for room heating, greenhouse heating and industrial drying, etc. In the current state-of-the-art, the maximum temperature for hot storage can go as high as 850°C in a concentrated solar thermal power plant. Cold thermal energy storage is used to provide house cooling, or cooling of food, vegetables, and fruit for storage and transportation. For example, cold packs or cold mass are increasingly commercially available for temporary house cooling, or for keeping food cold during transportation or for a short period of storage.

From the viewpoint of the materials used for thermal energy storage, there is sensible energy storage (with no state change in the material), latent heat storage (using phase change materials), or thermochemical energy storage, which involves cyclic chemical change from an original material to a midstage material and back to the original material. In a thermochemical energy storage process, thermal energy is absorbed or released during cyclic chemical reactions/changes of the materials. With the midstage material stored, the energy is stored. Therefore, long-term energy storage and a large storage capacity are possible using thermochemical energy storage.

1.3 EXAMPLES OF INDUSTRIAL APPLICATIONS OF THERMAL ENERGY STORAGE

Thermal energy storage (hot) has been widely applied in concentrating solar thermal power (CSP) plants for solar thermal energy storage during the day, to provide needed heat during the night to generate electrical power for 4–6 h. Solar thermal energy storage offers the advantage of a CSP plant against photovoltaic-based solar power generation. Therefore, almost every commercial CSP plant worldwide has a solar thermal storage system. Details including electrical power supply, power system thermal efficiency, the area of solar field, heat transfer fluid, thermal storage media, cooling mode, etc., for worldwide power plants and projects of concentrating solar power are available at the website of the U.S. National Renewable Energy Laboratory [2].

Ice or cold water storage is another thermal storage technology for energy conservation in industrial applications [3]. A large percentage of electricity (50%–75%) goes into buildings, and much of that runs air conditioning, either cooling or heating. To cope with the peak loads, particularly in summer, ice storage has been widely used. Typically, ice is made at night, when electricity is cheaper and it is cooler; while ice-melting provides cooling for air conditioning during the day when it is hot and electricity is in short supply. This can shift the peak of the demand curve and significantly reduce the level of peak power. When wind energy is considered, it is typical that the wind blows stronger at night than in the day and wind power electricity may be used to make ice for air conditioning in the day.

In conclusion, with the increasing demand and applications of energy storage in industry, the underlying physics and various engineering issues for thermal energy storage systems should be given a systematic study and explanation. To provide a comprehensive reference for readers (researchers and engineers) to better understand thermal energy storage technologies, this book focuses on several important issues:

(1) An overview of thermal energy storage requirements in industry, and the structural and configuration designs of thermal energy storage systems, and management of operation of thermal energy charge and discharge.

(2) Issues to be considered for selection of thermal storage materials and heat transfer fluid for high round-trip energy storage efficiency, from thermal energy back to thermal energy.

(3) Tools and modeling for quantitative calculation and analysis of energy transport and heat transfer between energy storage materials and energy-carrying fluid in thermal energy storage systems.

(4) Sizing of thermal (hot or cold) storage systems for required amount of thermal energy (hot or cold) storage in various industrial applications.

(5) Thermal protection and insulation for large-scale thermal storage systems.

(6) Mechanical issues of thermal storage devices due to thermal cycles of energy charge and discharge.

At the end of Chapter 4, following the presentation of the mathematical models and numerical solution method for thermal storage process, computer code is provided for the convenience of basic engineering computation and analysis of thermal storage systems and sizing of storage volumes.

REFERENCES

[1] Li PW. Energy storage is the core of renewable energy technologies. IEEE Nanotechnol Mag 2008;2:13–8.
[2] Website by U.S. National Renewable Energy Laboratory. http://www.nrel.gov/csp/solarpaces/by_country.cfm.
[3] Website of IceEnergy. https://www.ice-energy.com/company/.

CHAPTER 2

Thermal Storage System Configurations and Basic Operation

Contents

Abstract

A thermal storage process typically involves energy collecting, storage, and delivery using a heat transfer fluid (HTF). We can use the HTF alone as a thermal storage medium, or use dual media (with HTF and another material in a packed bed). System operation, configurations/arrangement of all possible packed-bed and fluid flow for the single medium or dual media thermal storage systems are presented and discussed in this chapter.

Keywords: Heat transfer fluid, Ideal thermal storage, Packed bed thermal storage, Phase change material, Thermal storage system, Thermocline thermal storage

2.1 THE SCENARIO OF IDEAL THERMAL STORAGE (HEAT OR COLD)

As the goal of thermal energy storage, it is always desirable that the stored "heat" or "cold" be delivered with no degradation of temperature [1]. A decrease of the temperature of "hot" thermal energy introduces a loss of exergy, or a reduction in efficiency, when the thermal energy is converted to mechanical or electrical energy. Increase in the temperature of a "cold"

http://dx.doi.org/10.1016/B978-0-12-805344-7.00002-X

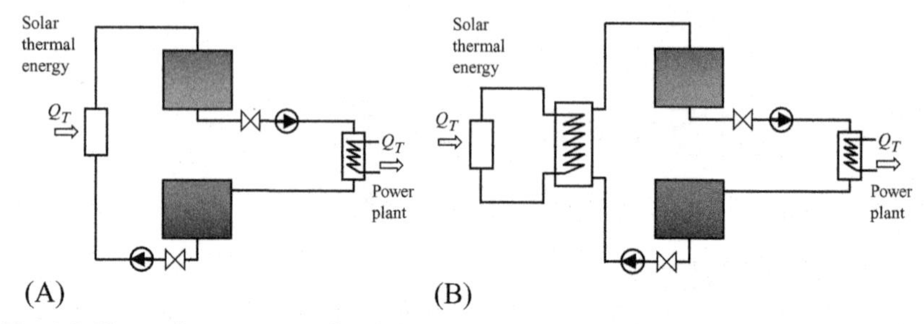

Fig. 2.1 Thermal energy stored in HTF using two tanks [2]. (A) The heat transfer fluid in solar collection system is also used for thermal storage. (B) The heat transfer fluid in solar heat collection system is different from the thermal storage fluid.

thermal storage material will cause extra electrical energy to be expended to drive the refrigeration system to compensate for the loss of cold.

To avoid degradation of temperatures of thermal energy, the best-case scenario is that the energy-carrying fluid in the thermal storage system, called the heat transfer fluid (HTF), be withdrawn at the same temperature at which it was originally stored. Therefore the ideal situation for the most efficient thermal energy storage system is to include two storage tanks for a heat transfer fluid (HTF), as shown in Fig. 2.1A. The thermal energy–carrying hot fluid is stored in a hot tank. To withdraw the energy, the fluid is pumped out, and after heat exchange, the fluid temperature is lowered and then stored in a cold tank. When there is a need to store thermal energy, the fluid from the cold tank is pumped out to be heated up and then stored in the hot tank. Assuming that the tanks are very well thermally insulated, the round-trip energy storage efficiency can approach almost 100%, meaning that the stored thermal energy temperature will not be degraded.

When implementing this energy storage and delivery system in a concentrated solar thermal power (CSP) plant, the thermal energy for storage is from solar concentrators [2]. The power generation system uses the stored thermal energy at times when sunlight is not available, whether at night or during bad weather.

A different version of thermal storage using fluid only is shown in Fig. 2.1B, where the fluids for thermal storage and HTF are different. One example, which is found in some concentrated solar power plants, uses parabolic trough collectors; synthetic oils are used as the HTF, while molten nitrate salts are used as thermal storage fluids. This arrangement is due to the different features of the two types of fluids; in particular, the HTF may not be suitable for thermal storage, either because of the cost or due to inadequate properties, such as density and heat capacity being too low.

Finally, it should be pointed out that the flow direction of the HTF during the heat discharging process is inverse to that of the heat charging process. This feature of flow direction inverse between heat charging and discharging processes is generally true, which will also be explained later in discussions for all other thermal storage systems.

2.2 THERMOCLINE THERMAL STORAGE SYSTEM USING HTF ALONE

Although its energy storage efficiency is ideal, the two-tank thermal energy storage system shown in Fig. 2.1 always keeps one tank space empty, which to some extent is a waste of tank space and thus is not cost-effective. Therefore, a so-called thermocline thermal storage system has been proposed, using only a single tank. There have been two types of designs using a single tank and only the HTF as the thermal storage material, as shown in Fig. 2.2. The operation of these two types of thermocline thermal storage requires that the hot fluid charges into the tank always from the top, while during hot fluid delivery, the flow direction is reversed so that cold fluid flows into the tank from the bottom and thus hot fluid is discharged out from the top. As a result of this operation, during both energy storage and delivery processes, hot fluid is always kept on top and cold always at the bottom, which creates hot-cold stratification due to the temperature gradient and buoyancy effect. This phenomenon is called a thermocline, and this technique has become well-accepted by the thermal storage community.

In Fig. 2.2B a thermal insulation baffle is used between the hot and cold fluids to separate the hot from mixing with cold, which makes the hot-cold separation or stratification even more ideal [3]. The concept of separating

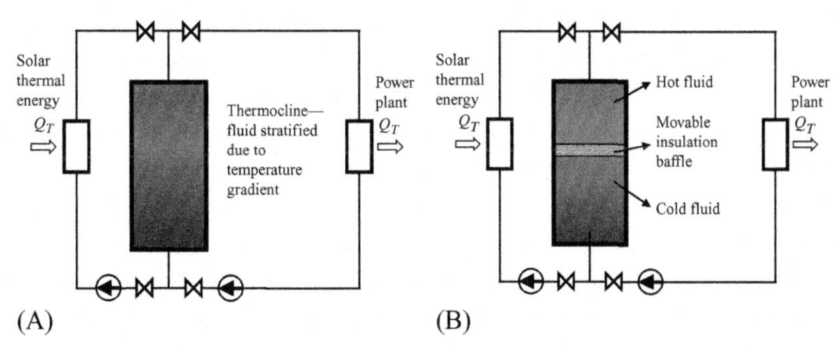

Fig. 2.2 Thermocline thermal storage system using a heat transfer fluid only. (A) Thermocline; (B) hot and cold fluid separated/insulated with a movable baffle.

hot from cold HTF using a movable thermal insulation baffle in a single tank is not necessarily limited to the vertical style.

Although the benefit of cost reduction using only one tank in thermocline thermal storage is attractive, destruction of the temperature stratification due to flow disturbance is a difficult issue for the system shown in Fig. 2.2A [4]. In the case of Fig. 2.2B, the proper separation of the hot and cold fluid using the baffle requires a moving and control mechanism, which can be inconvenient. Therefore, the systems in Fig. 2.2 have not been adopted in large-scale thermal storage systems in CSP industry so far.

2.3 THERMOCLINE THERMAL STORAGE SYSTEM WITH PACKED BED AND A HEAT TRANSFER FLUID

It also needs to be pointed out that, under certain circumstances, the HTF itself may not be suitable to be used as a thermal storage material [5], particularly if the HTF is too expensive, or the energy storage capacity ($\rho \cdot Cp$) of the HTF is not sufficiently high. Or, if the HTF needs a very high pressure to be kept in the liquid state, a large storage tank will suffer from large mechanical stresses, which will drive the cost of the thermal storage tank higher. Therefore, thermal energy storage with HTF flowing through a packed bed of another thermal storage material is also frequently used. In this case, the second thermal storage material must have much higher energy storage capacity ($\rho \cdot Cp$) for sensible energy storage, or a high value of ($\rho \cdot \Gamma$) for latent heat energy storage. It is noted here that the latent heat is represented by Γ. Even when a second material is used in a packed-bed thermal storage system, the system still has a certain amount of HTF inside which still stores a certain amount of energy. Therefore, this mode of energy storage is also called dual-media thermal storage.

The packed bed may be a sensible thermal storage material or phase change material (PCM). The general goal of using a packed bed with HTF fluid flowing through is to use low-cost materials, or to reduce the storage volume for a large quantity of thermal storage [6]. Heat transfer between the heat-carrying HTF and the other thermal storage material is inevitable in this case, such as is seen in Fig. 2.3. This means that when the thermal energy from the HTF is stored in another medium, a temperature decrease is inevitable and the energy storage efficiency cannot reach 100%. As the result, the discharged HTF from the thermal storage tank will suffer a degradation of the temperature after a certain time.

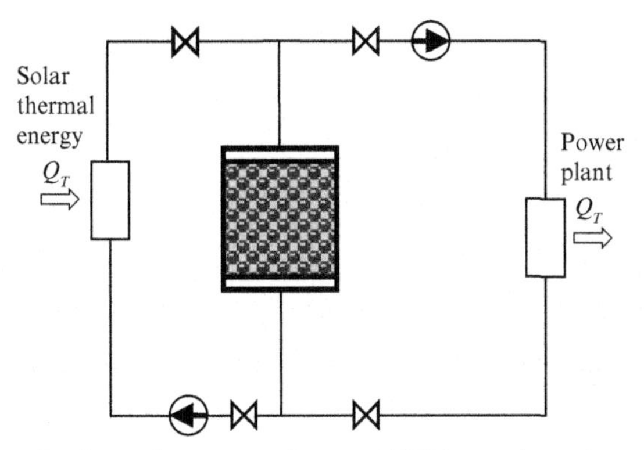

Fig. 2.3 Dull media thermal storage, where an HTF flows through a packed bed of another thermal storage material.

Similarly, the HTF flow directions in a dual-media thermal storage system also need to be inversed between the energy charging and discharging processes. A brief proof of this arrangement is given here, as illustrated in Fig. 2.4. Assuming that the HTF is charged into the tank from the top, it will result in a final temperature distribution in the tank as indicated by the "after-charge" curve. If cold fluid flows in from the top of the tank to take the stored heat out, its temperature variation is indicated by the curve of the discharge scheme (b). It can be seen that the cold fluid temperature will reach a point equal to that of the thermal storage material as it passes through the storage tank. After that point, the HTF has to release its heat

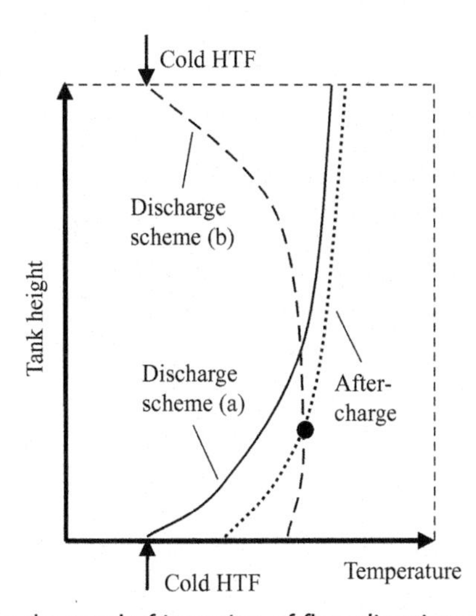

Fig. 2.4 Illustration for the need of inversion of flow direction of HTF.

to the thermal storage material, which is now colder than the fluid as the HTF continues flowing through the remainder of the storage tank. If the cold HTF enters in the tank from the bottom to extract stored heat, its temperature variation will be as illustrated by the curve of the discharge scheme (a). In this case the temperature of the HTF will keep rising throughout its path to the top end of the storage tank.

One needs to check and manage the convective heat transfer between the HTF and the material of the packed bed in order to achieve the best energy storage efficiency. The Newton cooling equation for heat transfer rate is

$$\dot{Q} = Ah\left(T_f - T_s\right) \tag{2.1}$$

It is understandable from this equation that for the HTF to give energy to a solid material with minimum temperature degradation, both a large heat transfer area between the fluid and the solid and a large convective heat transfer coefficient h are necessary. For better accommodation of this requirement, the advantages and disadvantages of the packaging schemes of thermal storage materials should be evaluated before doing the system design and engineering.

There are four typical schemes for thermal storage material packaging, as shown in Fig. 2.5. These are: (A) packed rocks/pebbles of sensible thermal storage materials; (B) packed capsules which have PCM encapsulated inside; (C) multiple zones with different types of capsules having different phase change temperatures, or even a hybrid package in which rocks/pebbles of

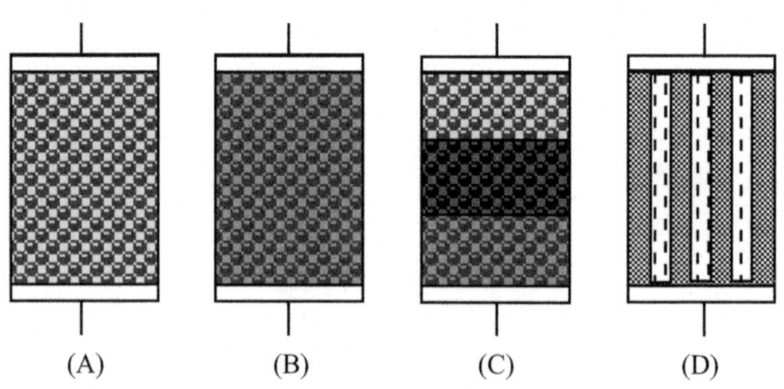

Fig. 2.5 Packaging schemes of packed-bed thermal storage. (A) Rocks or pebbles in packed bed; (B) PCM capsules in packed bed; (C) multizones of PCM capsules and rocks/pebbles packed bed; (D) "block" thermal storage material (sensible or PCM) with HTF pipes embedded inside.

sensible thermal storage materials are on top and capsules of PCMs are packed underneath; (D) heat transfer fluid flowing in pipes that are embedded in a "block" of thermal storage material, either sensible thermal storage material or PCM. Detailed discussions about these packaging schemes are provided in the following sections.

2.3.1 Packed Bed Using Pebbles of Solid Material for Sensible Energy Storage

Sensible thermal energy storage materials such as rocks, pebbles, sands, or fabricated concrete pebbles may be used to fill a packed bed, as shown in Fig. 2.5A. These materials are easy to obtain at a low cost and their thermal energy storage capacity $\rho \cdot Cp$ is also sufficiently high. Thermal/physical properties of these materials are given and discussed in Chapter 3. When the grain size (of pebbles, rocks, or sands) is approximately uniform, the porosity of the packed bed may range from 0.35 to 0.41 [7]. A packed bed with nonuniform grain size, such as rocks of different sizes, could have a porosity as low as 0.326 [8], and rocks mixed with sand could have a porosity even lower, around 0.22 [4].

Due to the very large surface area of the rocks, pebbles, or sands that form the porous media, the heat transfer between the HTF and the thermal storage material will be very good, so that the temperature difference $\left(T_f - T_s\right)$ can be rather small, which contributes to high thermal energy storage efficiency.

As schematically shown in the structure of the thermal storage tanks in Fig. 2.5, the top and bottom fluid chambers of the tank need to be designed carefully to ensure that the fluid is distributed uniformly in the radial direction and also that the packaging is uniformly placed, leaving no channeling in the porous media [8]. Fortunately, some studies have shown that the flow of HTF in a packed bed can reach uniformity in a relatively short length of the tank [9,10].

2.3.2 PCMs Encapsulated as Fillers for Packed Beds

Due to the relatively large latent heat (in the phase change process) compared to sensible heat, PCMs are attractive for large quantity thermal energy storage with small volume. This introduces another type of dual-media thermal storage with HTF flowing through a packed bed of PCM materials, as shown in Fig. 2.5B. Again, in order to have a large heat transfer area between the PCM and HTF for better heat transfer, PCM materials are better encapsulated into small capsules [11]. Technologies of encapsulation of PCMs have been significantly developed in recent years, particularly for low to medium thermal storage applications up to temperatures of 250°C. Encapsulation technologies for high temperature PCMs, above 500°C, are still under development [12,13].

2.3.3 PCMs and PCM-Sensible Hybrid Packed Bed Thermal Storage

It is known that during the energy charging process, the temperature of the HTF must be higher than that of the PCM. When thermal energy is discharged, the temperature of the HTF is lower than that of the PCM. This nature of a significantly lowered temperature of the HTF from charging to discharging is a disadvantage. To avoid any large degradation of the temperature of the HTF, multiple PCMs packed in multiple zones (as shown in Fig. 2.5C) with cascade melting points have been proposed and studied. It has been proven that the thermal storage efficiency can be improved using cascade PCMs packed in a single tank or multiple tanks [14,15].

To further expand the benefit of using cascade PCMs, sensible thermal storage materials may be placed on the top of the storage system [16].

2.3.4 Solid Block Thermal Storage With HTF Pipes Embedded

Other than the porous packed bed using particles, pebbles, or PCM capsules, thermal storage may also use integrated material, referred to as "blocks." Concrete [17,18], sands saturated by thermal conductive liquid [7,19], and "blocks" of PCM materials [20] have been used and reported for thermal energy storage. In this case, pipes of HTF are embedded into the thermal storage block material, as shown in Fig. 2.5D. The material can be either for sensible thermal storage or for latent heat thermal storage.

There are three types of configurations of HTF and solid block materials, as shown in Fig. 2.6. The first configuration uses plate type thermal storage blocks that form 2D flow channels. The second type has thermal storage material as rods and HTF flows in the gap of the bundle of the rods. The third type has a large block of thermal storage material with pipes of HTF embedded through it.

Unlike the case of the HTF flowing in a porous thermal storage material, the heat transfer surface area of the channels/pipes embedded in thermal storage material is much smaller. Therefore, enhancement of heat transfer in the solid block and at the surface of the channels/pipes is needed.

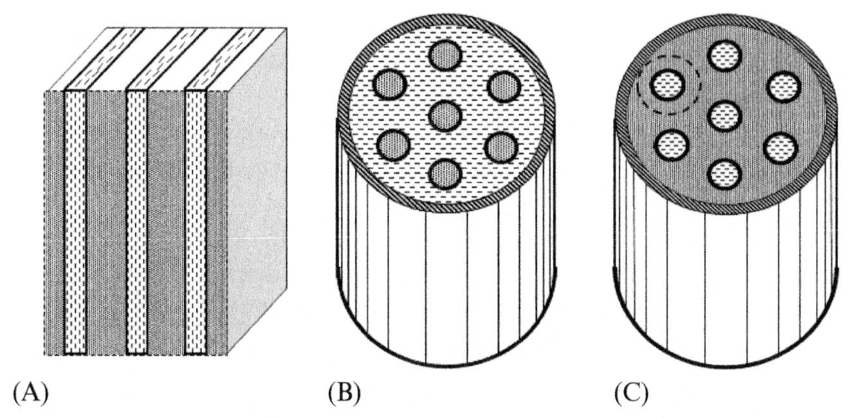

Fig. 2.6 Configurations of HTF channels in thermal storage block material. Areas shadowed with dashed lines show the region of the HTF. (A) Fluid flows in between solid slabs. (B) Fluid flows in between solid rods. (C) Fluid flows in pipes embedded in solid.

Measures such as putting metal foam and metal coils in the thermal storage material and fins at the outside of the pipes are typically recommended [20].

The thermophysical properties of the block thermal storage materials, such as concrete, sands saturated with thermal conductive liquid, PCM, and issues of heat transfer enhancement in these materials are presented in Chapter 3.

Convective heat transfer coefficients of HTF in a variety of porous media and in the flow channels as shown in Fig. 2.6 are discussed in Chapter 4, where heat transfer and energy storage performance are analyzed.

2.4 THERMOCHEMICAL ENERGY STORAGE TECHNOLOGIES

Although either sensible heat or latent heat thermal storage technologies make it possible to store a large amount of energy for electrical power generation, on the order of 300 MWe for 6–8 h, it is still very challenging to use these technologies for thermal energy storage for a much longer period of time, say a season or from summer to winter. The major challenge is the heat loss due to imperfect thermal insulation, or the high cost for highly reliable thermal insulation. Therefore, researchers have been looking for other technologies, such as thermochemical energy storage, to solve the demand for long-term thermal energy storage.

The basic idea of thermochemical energy storage is to use the heat to drive an endothermic chemical reaction and dissociate material A to material B and C ($A + heat \rightarrow B + C$). The materials B and C are able to be stored for a desired length of time and can even be transported to different locations. When desired at another time, one can use the reverse reaction ($B+C \rightarrow A+$heat), which is exothermic, to release heat for utilization. Depending on the chemical reaction chosen, thermochemical approaches can provide energy storage and return back heat both at low temperature (for room heating) or rather high temperatures, up to several hundred degrees Celsius.

Some reduction-oxidation (redox) reactions with significant heat effects have been surveyed in Ref. [21] for application in thermochemical energy storage. Metal oxides such as MnO_2 and Co_3O_4 may be redox cycled so that thermal energy storage and retrieval can be accomplished. When input with heat, the following reduction reactions may occur:

$$2MnO_2 + \text{heat} \rightarrow Mn_2O_3(s) + \tfrac{1}{2}O_2 \tag{2.2}$$

$$Co_3O_4 + \text{heat} \rightarrow 3CoO\,(s) + \tfrac{1}{2}O_2 \tag{2.3}$$

The oxidation reactions of Mn_2O_3 and CoO release heat and thus thermal energy is retrieved. These technologies are considered for large-scale thermal storage deployment in concentrated solar power (CSP) plants that use air as the heat transfer fluid. The convenience of using oxygen as the reactive gas is because of the supply of ambient air, which thus avoids storage of gaseous material. The redox temperature range of these reactions is suitable for new generation solar tower-based CSP plants, which have a temperature of 900°C under atmospheric pressure. For example, the reaction, Eq. (2.2), also has a high energy density of (844 kJ/kg Co_3O_4) and its performance was reported to be very stable during cycling reactions [22,23].

The cycling between metal hydroxide to metal oxide is also suitable to be applied for thermochemical energy storage. The reaction, such as:

$$Ca(OH)_2(s) + \text{heat} \rightarrow CaO(s) + H_2O(g) \tag{2.4}$$

has been reported in Ref. [23] for energy storage in a fluidized reactor.

Dehydration reactions can also be utilized for thermochemical energy storage; the following reaction:

$$CaCl_2 \cdot 6H_2O + \text{heat} \rightarrow CaCl_2 + 6H_2O \tag{2.5}$$

has been adopted for thermal storage at low temperature range.

Chemical reactions with gaseous materials such as the association-synthesis system:

$$2NH_3 + \text{heat} = N_2 + 3H_2 \tag{2.6}$$

has also been considered for thermochemical energy storage. Since this system uses catalysts and operates at rather high pressures (80–200 bar), it needs storage of H_2 and N_2 gases, which is not cost effective for large quantity thermal storage.

Some more complicated chemical processes have also been utilized for thermochemical energy storage. The following series reactions are examples to accomplish the goal of thermal energy storage:

$$2H_2SO_4 + \text{Heat} \rightarrow 2H_2O + 2SO_2 + O_2 \text{ (Sulfuric acid decomposition)} \tag{2.7}$$

$$2H_2O + 3SO_2 \rightarrow 2H_2SO_4 + S \text{ (disproportionation producing sulfur)} \tag{2.8}$$

$$S + O_2 \rightarrow SO_2 + \text{Heat (Combustion of sulfur for heat)} \tag{2.9}$$

Here SO_2 and S are recycled in the entire processes to achieve the goal of energy storage.

Thermochemical hydrogen production has also been considered for thermal energy storage in hydrogen fuel. One example of a thermochemical cycle for H_2 production based on CeO_2/Ce_2O_3 oxides was described and demonstrated in [24]. The thermochemical reaction consists of two steps:

$$\text{Reduction}: 2CeO_2 \rightarrow Ce_2O_3 + 0.5O_2;$$

$$\text{Hydrolysis}: Ce_2O_3 + H_2O \rightarrow 2CeO_2 + H_2.$$

The thermal reduction of Ce(IV) to Ce(III) (endothermic step) can be performed in a solar reactor under a controlled inert atmosphere. The first step reaction has been demonstrated at operating conditions of $T = 2000°C$, $P = 100$–200 mbar. The hydrogen generation step for water-splitting with Ce(III) oxide has been accomplished in a fixed-bed reactor and the reaction is complete with a fast kinetic in at a temperature range of 400–600°C. The recovered Ce(IV) oxide is then recycled to feed the first step. In the entire process, water is the material input and heat is the energy input. The output material is hydrogen fuel and oxygen. Importantly, these two gases are obtained in different steps that avoid a challenge of high temperature gas-phase separation. Pure hydrogen is produced that can be stored as fuel. Although the operation at a temperature of 2000°C faces a big material challenge, the cerium oxide two-step thermochemical is still a promising process for hydrogen production.

Details of heat values (additions and retrieves) involved in the previously discussed chemical reactions and the devices accommodating the reactions and heat addition/retrieves studied by various research groups/researchers are introduced in Chapter 4.

REFERENCES

[1] Li P, Van Lew J, Karaki W, Chan C, Stephens J, O'Brien JE. Transient heat transfer and energy transport in packed bed thermal storage systems. In: dos Santos Bernardes MA, editor. Developments in heat transfer. InTech; 2011. ISBN: 978-953-307-569-3 [chapter 20, First published, August].
[2] Stekli J, Irwin L, Pitchuccmani R. Technical challenges and opportunities for concentrating solar power with thermal energy storage. ASME J Therm Sci Eng Appl 2013;5:021011-1.
[3] Copeland RJ. US patent #4523629A. June 18, 1985.
[4] Pacheco JE, Showalter SK, Kolb WJ. Development of a molten-salt thermocline thermal storage system for parabolic trough plants. J Sol Energy Eng 2002;124(2):153–9.
[5] Kearney D, Herrmann U, Nava P, Kelly B, Mahoney R, Pacheco J, et al. Assessment of a molten salt heat transfer fluid in a parabolic trough solar field. J Sol Energy Eng 2003;125(2):170–6.
[6] Brosseau D, Kelton JW, Ray D, Edgar M, Chrisman K, Emms B. Testing of thermocline filler materials and molten-salt heat transfer fluids for thermal energy storage systems in parabolic trough power plants. J Sol Energy Eng 2005;127(1):109–16.
[7] Han BX, Li P, Kumar A, Yang Y. Experimental study of a novel thermal storage system using sands with high-conductive fluids occupying the pores. In: IMECE 2014-38999, proceedings, ASME 2014 international mechanical engineering congress & exposition, Montreal, Quebec, Canada, November 14–20; 2014.
[8] Valmiki MM, Karaki W, Li P, Van Lew J, Chan C, Stephens J. Experimental investigation of thermal storage processes in a thermocline tank. J Solar Energy Eng 2012;134:041003.
[9] Gabbrielli R, Zamparelli C. Optimal design of a molten salt thermal storage tank for parabolic trough solar power plants. J Solar Energy Eng 2009;131:041001.
[10] Yang Z, Garimella SV. Molten-salt thermal energy storage in thermocline under different environmental boundary conditions. Appl Energy 2010;87(11):3322e9.
[11] Xu B, Li P, Chan C. Application of phase change materials for thermal energy storage in concentrated solar thermal power plants: a review to recent developments. Appl Energy 2015;160:286–307.
[12] Nomura T, Zhu C, Sheng N, Saito G, Akiyama T. Microencapsulation of metal-based phase change material for high-temperature thermal energy storage. Sci Rep 2015;5:9117. http://dx.doi.org/10.1038/srep09117.
[13] Zhao CY, Zhang GH. Review on microencapsulated phase change materials (MEPCMs): fabrication, characterization and applications. Renew Sust Energ Rev 2011;15(8):3813–32.
[14] Galione PA, Pérez-Segarra CD, Rodríguez I, Lehmkuhl O, Rigola J. A new thermocline-PCM thermal storage concept for CSP plants, Numerical analysis and perspectives. Energy Procedia 2014;49:790–9.
[15] Tumilowicz E, Chan CL, Li P, Xu B. An enthalpy formulation for thermocline with encapsulated PCM thermal storage and benchmark solution using the method of characteristics. Int J Heat Mass Transf 2014;79:362–77.
[16] Nallusamy N, Sampath S, Velraj R. Experimental investigation on a combined sensible and latent heat storage system integrated with constant/varying (solar) heat sources. Renew Energy 2007;32:1206–27.
[17] Wu M, Li MJ, Xu C, He YL, Tao WQ. The impact of concrete structure on the thermal performance of the dual-media thermocline thermal storage tank using concrete as the solid medium. Appl Energy 2014;113:1363–71.
[18] John E, Hale M, Selvam P. Concrete as a thermal energy storage medium for thermocline solar energy storage systems. Sol Energy 2013;96:194–204.

[19] Yongping YANG, Jingxiao HAN, Peiwen LI, Ben XU, Hongjuan HOU. Thermal energy storage characteristics of synthetic oil and sand mixture for thermocline single tank. Proc CSEE 2015;35(3) [in Chinese].
[20] Laing D, Bauer T, Steinmann WD, Lehmann D. Advanced high temperature latent heat storage system—design and test results. In: Proceedings of the 11th international conference on thermal energy storage—Effstock, June; 2009. p. 14–7.
[21] Tescaria S, Lantin G, Lange M, Breuer S, Agrafiotis C, Roeb M, et al. Numerical model to design a thermochemical storage system for solar power plant. Energy Procedia 2015;75:2137–43.
[22] Tescari S, Agrafiotis C, Breuer S, de Oliveira L, Puttkamer M, Roeb M, et al. Thermochemical solar energy storage via redox oxides: materials and reactor/heat exchanger concepts. Energy Procedia 2014;49:1034–43.
[23] Pardo P, Anxionnaz-Minvielle Z, Rouge S, Cognet P, Cabassud M. $Ca(OH)_2/CaO$ reversible reaction in a fluidized bed reactor for thermochemical heat storage. Sol Energy 2014;107:605–16.
[24] Abanades S, Flamant G. Thermochemical hydrogen production from a two-step solar-driven water-splitting cycle based on cerium oxides. Sol Energy 2006;80(12):1611–23.

CHAPTER 3
Thermal Energy Storage Materials

Contents

Abstract

Materials needed for thermal storage include heat transfer fluids (HTFs), fluids for sensible energy storage, solids and solid-fluid mixtures for sensible energy storage, and also materials for latent heat thermal storage. This chapter will first present the requirements for materials used for thermal energy storage including an HTF and solid materials. Because of the trend in which fluids are becoming popularly used for the HTF and the liquid-alone thermal storage material, some currently available HTFs and their properties are provided for engineering application. To support further development for high-temperature HTFs for both heat transfer and thermal storage application, criteria for selection of the fluids based on their thermal and transport properties are discussed. At the end of the chapter, properties of some solid materials for sensible and latent thermal storage applications are also provided.

Keywords: Entropy production, Heat transfer fluid, Phase change material, Thermal energy storage, Thermal storage media

Nomenclature

- ***A*** total heat transfer area in a heat exchanger or a heat receiver (m^2)
- ***Cp*** heat capacity [J/(kg K)]
- ***d*** inner diameter of pipes with fluid flow (m)
- ***D_h*** hydraulic diameter of fluid pipes (m)
- ***f*** Darcy friction coefficient
- ***h*** heat transfer coefficient [W/(m^2K)]

http://dx.doi.org/10.1016/B978-0-12-805344-7.00003-1

k thermal conductivity of fluid [W/(m K)]
L length of a fluid pipe (m)
$\dot{m}$ mass flow rate of heat transfer fluid (kg/s)
n number of fluid tubes
Pr Prandtl number
$\dot{Q}$ heat transfer rate or thermal power (W)
Re Reynolds number
s specific entropy [J/(kg K)]
$\dot{S}_{gen}$ total entropy generation rate [J/(s K)]
T temperature (K)
U velocity (m/s)
v specific volume of fluid (m^3/kg)
$\dot{V}$ volumetric flow rate of fluid (m^3/s)

Greek symbols

ΔP pressure loss in pipe flow (Pa)
μ dynamic viscosity (Pa s)
ρ density (kg/m^3)

Subscripts

1, 2 locations
c circulation loop
collector solar thermal collector
exchanger heat exchanger
gen generation rate
h high temperature
h-w high temperature at wall
HE heat exchanger
HE-h high temperature heat exchanger
HE-l low temperature heat exchanger
i, in inlet
l low temperature
l-w low temperature at wall
m average temperature
o, out outlet

Superscripts

Laminar due to laminar flow
T-1 turbulent at $2300 \leq \text{Re} \leq 2 \times 10^4$
T-2 turbulent at $\text{Re} \geq 2 \times 10^4$
Turbulent due to turbulent flow

3.1 THERMAL STORAGE ASSOCIATED WITH DIFFERENT TYPES OF MEDIA

Thermal energy storage can be accomplished by directly storing a hot heat transfer fluid (HTF) in a tank and then withdrawing it, which theoretically could maintain a round-trip efficiency of 1.0, if the thermal insulation could be made perfect. For a relatively low cost system, the fluid should have a reasonably high density and heat capacity in order to store a sufficiently large amount of energy in a relatively small volume. Therefore, a larger product of density and heat capacity $(\rho \cdot C)$ is a basic requirement for thermal storage materials. This is true for both solid and liquid materials for thermal storage. In cases where phase change material is concerned, this basic requirement becomes the product of density and heat of fusion $(\rho \cdot \Gamma)$.

For liquid-alone thermal storage, the stored hot fluid (at a temperature of T_H) can deliver energy at a flow rate of $\dot{m}$ and a temperature drop of $(T_H - T_L)$ for a desired thermal power of $\dot{Q}_T$. The thermal power delivery can be expressed as:

$$\dot{Q}_T = \dot{m} \cdot C_f (T_H - T_L) \tag{3.1}$$

If the time period of thermal power delivery is Δt, the total energy and the volume for fluid storage are given in the form of:

$$Q_T = \Delta t \cdot \dot{Q}_T \tag{3.2}$$

$$V_{f-ideal} = \Delta t \cdot \dot{m} / \rho_f \tag{3.3}$$

There are two possible reasons to use solid materials to form a packed bed for thermal energy storage rather than using a fluid directly:

(1) The density and heat capacity of the fluid are both low so that $V_{f-ideal}$ could be extremely large.

(2) The fluid is too expensive and therefore relatively cheaper solid materials (in a packed bed) are used for primary thermal storage. The fluid's role is mainly for heat transfer and at the same time it has a limited amount of energy stored when it fills the voids (ε) in the packed bed.

In cases where a solid medium is mainly used for the thermal storage, with a fluid flowing through to carry heat in and out, the storage tank volume, $V_{\tan k}$, must satisfy the following condition [1]:

$$V_{\tan k}\left[\varepsilon\left(\rho_f C_f\right) + (1-\varepsilon)(\rho_s C_s)\right] > V_{f-ideal}\left(\rho_f C_f\right) \tag{3.4}$$

Here, $V_{\tan k}$ is the volume of the storage system packed with both solid and fluid material, which can be larger or smaller than $V_{f-ideal}$, depending on the situations as discussed here. When $(\rho_s C_s)$ is much larger than $(\rho_f C_f)$, $V_{\tan k}$ can be smaller than $V_{f-ideal}$. For many sensible heat based thermal storage systems with $(\rho_s C_s)$ not much larger than $(\rho_f C_f)$, $V_{\tan k}$ must be typically larger than $V_{f-ideal}$. For latent heat based thermal storage using a phase change material (PCM), the product of density and latent heat $(\rho_s \Gamma_s)$ of the PCM is typically much larger than $(\rho_f C_f)$, and therefore $V_{\tan k}$ can be smaller than $V_{f-ideal}$.

Due to the involvement of heat transfer in thermal storage processes, both the fluid and the solid materials must have favorable thermal and transport properties for both heat transfer and energy storage. Therefore, it is necessary to discuss the properties and criteria for fluid and solid materials that can best serve the thermal storage purpose.

3.2 FLUIDS FOR HEAT TRANSFER AND THERMAL ENERGY STORAGE APPLICATIONS

Fluids capable of operating at high temperatures with low vapor pressure are commonly used in solar thermal systems to transfer heat from the concentrator (heat source) to a destination for storage or delivering the heat to a working fluid in a power plant, as shown in Fig. 3.1. Fluids may also be used for both heat transport (HTFs) and thermal energy storage material at the

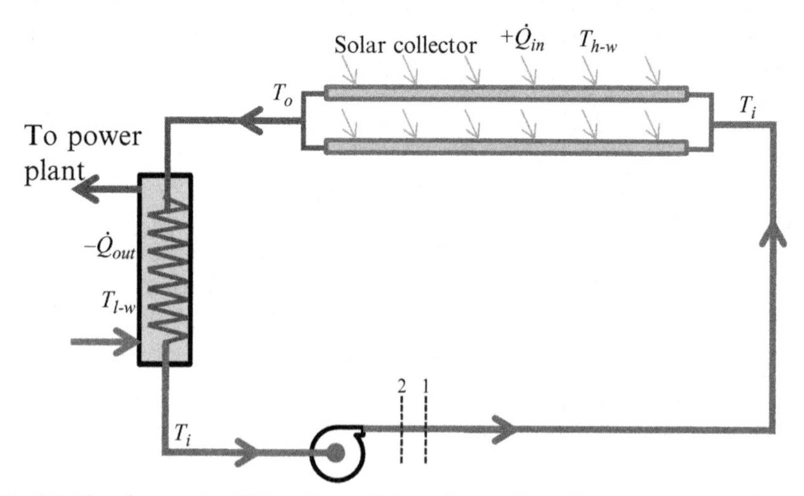

Fig. 3.1 Flow loop using HTF to transmit heat from solar collectors to power plant.

same time. For thermal storage in a compact volume, a large heat capacity and density are preferable. Good transport properties of the fluid materials are important, both for the heat transfer purpose and for efficient thermal storage applications.

For a high efficiency thermal-to-electrical energy conversion in a concentrated solar thermal power (CSP) plant, it is advantageous to have a heat source/reservoir temperature as high as possible. While modern solar concentrator technology has made it possible to have a high temperature at the solar receiver, it has been a challenge to find an HTF that possesses all the desirable thermal and transport properties at high temperatures suitable for transporting heat to the thermal power system for power generation. Because of this challenge, the thermal efficiency of CSP plants is largely restricted by the highest operating temperature of the HTFs [2,3].

A variety of fluids can be used as heat transfer fluids, such as air, water, various types of oils, molten salts, and liquid metals. However, for high-temperature heat transfer at a temperature level well above 150°C, with the requirement of relatively low vapor pressure in the fluid and low corrosion in pipes and containers at the high temperatures, the fluid options are relatively few. When the fluid is also considered for thermal energy storage, the number of candidate fluids becomes even less.

So far, a limited number of fluids have been applied in the CSP industry for use in concentrated solar thermal power systems for both HTFs and thermal storage materials. The maximum working temperatures of these HTFs range from 300°C to 800°C.

Nontoxic petroleum-based oils (such as XCELTHERM 600, a C_{20} paraffin oil from Radco Industries) are representative HTFs from the early stages of CSP technology. XCELTHERM 600 has an upper-limit temperature of 316°C and a low temperature of below 0.0°C [3,4]. A demonstrative 1.0 MWe CSP plant (operated by Arizona Public Service) in the United States used XCELTHERM 600 as its HTF for their parabolic trough concentrators. However, due to the limited maximum temperature, XCELTHERM 600 has not been used as an HTF for advanced CSP systems, either for large-scale parabolic trough or solar power tower systems. Nevertheless, for solar thermal collection and thermal storage for applications such as cooking, drying, and house heating, XCELTHERM 600 is a benign HTF [5]. Its transport properties are given in Table A.1.

Therminol VP-1 (Eastman) is a type of synthetic oil which is a mixture of diphenyl oxide and biphenyl in a weight percentage of 73.5% vs. 26.5%. The oil has a freezing point of 12°C and an upper limit temperature of 400°C, at

Table A.1 Properties of XCELTHERM 600 in its liquid state [3]

Temperature (°C)	Dynamic viscosity 10^{-3} (Pa s)	Density (kg/m^3)	Heat capacity (kJ/(kg°C))	Thermal conductivity (W/(m°C))	Vapor pressure (Pa)
10.00	75.697	857.8	1.96	0.1369	–
15.60	50.182	854.4	1.98	0.1364	–
18.30	41.886	852.7	1.99	0.1362	–
21.10	35.431	851.1	1.99	0.1359	–
26.70	26.191	847.7	2.01	0.1355	–
32.20	20.055	844.4	2.03	0.135	–
37.80	15.489	841	2.05	0.1347	–
43.30	12.087	837.6	2.08	0.1342	–
48.90	9.734	834.3	2.1	0.1338	–
54.40	7.973	830.9	2.12	0.1333	–
60.00	6.625	827.5	2.13	0.1329	–
65.60	5.574	824.1	2.15	0.1324	–
71.10	4.742	820.8	2.17	0.132	–
76.70	4.072	817.4	2.19	0.1315	–
82.20	3.526	814.1	2.21	0.1311	–
87.80	3.077	810.7	2.23	0.1306	–
93.30	2.703	807.3	2.25	0.1302	9.8
98.90	2.432	804	2.27	0.1297	9.8
104.40	2.191	800.6	2.29	0.1293	9.8
110.00	1.983	797.2	2.3	0.1288	19.6
115.60	1.801	793.8	2.32	0.1284	29.4
121.10	1.642	790.5	2.34	0.1279	39.2
126.70	1.503	787.1	2.36	0.1275	49

132.20	1.382	785.4	2.38	0.127	68.6
137.80	1.27	780.3	2.4	0.1266	88.2
143.30	1.172	776.9	2.42	0.1261	107.8
148.90	1.085	773.6	2.44	0.1257	147
154.40	1.01	770.2	2.46	0.1252	186.2
160.00	0.945	766.8	2.47	0.1248	235.2
165.60	0.885	763.4	2.49	0.1243	294
171.10	0.832	760.1	2.51	0.1239	362.6
176.70	0.783	756.7	2.53	0.1234	450.8
182.20	0.737	753.3	2.55	0.123	558.6
187.80	0.696	749.9	2.57	0.1225	686
193.30	0.658	746.6	2.59	0.1221	833
198.90	0.623	743.2	2.61	0.1216	1009.4
204.40	0.59	739.8	2.62	0.1212	1215.2
210.00	0.559	736.4	2.64	0.1207	1460.2
215.60	0.532	733	2.66	0.1203	1744.4
221.10	0.507	729.7	2.68	0.1198	2077.6
226.70	0.483	726.3	2.7	0.1194	2459.8
232.20	0.46	722.9	2.72	0.1189	2910.6
237.80	0.439	719.5	2.74	0.1185	3420.2
243.30	0.42	716.2	2.76	0.118	4008.2
248.90	0.402	712.8	2.78	0.1176	4694.2
254.40	0.385	709.4	2.79	0.1171	5458.6
260.00	0.368	706	2.81	0.1167	6340.6
262.80	0.358	704.3	2.82	0.1164	6830.6
265.60	0.351	702.6	2.83	0.1162	7350

Continued

Table A.1 Properties of XCELTHERM 600 in its liquid state [3]—cont'd

Temperature (°C)	Dynamic viscosity 10^{-3} (Pa s)	Density (kg/m^3)	Heat capacity (kJ/(kg°C))	Thermal conductivity (W/(m°C))	Vapor pressure (Pa)
271.10	0.337	699.3	2.85	0.1158	8486.8
276.70	0.325	695.9	2.87	0.1153	9770.6
282.20	0.313	692.5	2.89	0.1149	11,221
287.80	0.301	689.2	2.91	0.1144	12,857.6
293.30	0.29	685.9	2.93	0.114	14,690.2
298.90	0.281	682.5	2.94	0.1135	16,748.2
304.40	0.27	679.1	2.96	0.1131	19,051.2
310.00	0.261	675.7	2.98	0.1126	21,618.8
315.60	0.252	672.4	3	0.1122	24,490.2

which the vapor pressure of the fluid is around 10 atm. Most large-scale parabolic trough concentrated solar thermal power plants use Therminol VP-1 for the HTF. Because of the relatively greater cost, Therminol VP-1 could not be used as a thermal storage material in parabolic trough-based CSP plants; instead, nitrate molten salts (such as HITEC salt) are widely used as a thermal storage liquid for the benefit of its relatively lower cost [6,7] and also low vapor pressure. Since the thermal energy storage material is different from an HTF, a heat exchanger must be used in this case. Thermal and transport properties of Therminol VP-1 are given in Table A.2.

Table A.2 Properties for Therminol VP-1 in its liquid state[a] [7]

Temperature (°C)	Dynamic viscosity 10^{-3} (Pa s)	Density (kg/m^3)	Heat capacity (kJ/(kg°C))	Thermal conductivity (W/(m°C))	Vapor pressure (kPa)
12	5.48	1071	1.523	0.137	–
20	4.29	1064	1.546	0.136	–
30	3.28	1056	1.575	0.135	–
40	2.6	1048	1.604	0.134	–
50	2.12	1040	1.633	0.133	–
60	1.761	1032	1.662	0.132	–
70	1.492	1024	1.69	0.131	–
80	1.284	1015	1.719	0.13	–
90	1.119	1007	1.747	0.129	–
100	0.985	999	1.775	0.128	0.5
110	0.875	991	1.803	0.126	0.8
120	0.784	982	1.831	0.125	1
130	0.707	974	1.858	0.124	2
140	0.642	965	1.886	0.123	3
150	0.585	957	1.913	0.121	5
160	0.537	948	1.94	0.12	7
170	0.494	940	1.968	0.118	9
180	0.457	931	1.995	0.117	13
190	0.424	922	2.021	0.115	18
200	0.395	913	2.048	0.114	24
210	0.368	904	2.075	0.112	32
220	0.345	895	2.101	0.111	42
230	0.324	886	2.128	0.109	54
240	0.305	877	2.154	0.107	68
250	0.288	867	2.181	0.106	86
260	0.272	857	2.207	0.104	108
270	0.258	848	2.234	0.102	133
280	0.244	838	2.26	0.1	163

Continued

Table A.2 Properties for Therminol VP-1 in its liquid state[a] [7]—cont'd

Temperature (°C)	Dynamic viscosity 10^{-3} (Pa s)	Density (kg/m^3)	Heat capacity (kJ/(kg°C))	Thermal conductivity (W/(m°C))	Vapor pressure (kPa)
290	0.232	828	2.287	0.098	198
300	0.221	817	2.314	0.096	239
310	0.211	806	2.341	0.095	286
320	0.202	796	2.369	0.093	340
330	0.193	784	2.397	0.091	401
340	0.185	773	2.425	0.089	470
350	0.177	761	2.454	0.086	548
360	0.17	749	2.485	0.084	635
370	0.164	736	2.517	0.082	732
380	0.158	723	2.551	0.08	840
390	0.152	709	2.588	0.078	959
400	0.146	694	2.628	0.076	1090
410	0.141	679	2.674	0.073	1230
420	0.137	662	2.729	0.071	1390
425	0.134	654	2.76	0.07	1470

[a]The properties are also expressed by the following correlations.

$\rho\ (\mathrm{kg/m^3}) = 1083.25 - 0.90797 \cdot T(°\mathrm{C}) + 0.00078116 \cdot T^2(°\mathrm{C}) - 2.367 \times 10^{-6} \cdot T^3(°\mathrm{C})$

$Cp(\mathrm{kJ/kg°C}) = 1.498 + 0.002414 \cdot T(°\mathrm{C}) + 5.9591 \times 10^{-6} \cdot T^2(°\mathrm{C}) - 2.9879 \times 10^{-8} \cdot T^3(°\mathrm{C}) + 4.4172 \times 10^{-11} \cdot T^4(°\mathrm{C})$

$\nu(\mathrm{m^2/s}) = 10^{-6} \cdot \exp\left(\frac{544.149}{T(°\mathrm{C}) + 114.43} - 2.59578\right)$

$k(\mathrm{W/m°C}) = 0.137743 - 8.19477 \times 10^{-5} \cdot T(°\mathrm{C}) - 1.92257 \times 10^{-7} \cdot T^2(°\mathrm{C}) + 2.5034 \times 10^{-11} \cdot T^3(°\mathrm{C}) - 7.2974 \times 10^{-15} \cdot T^4(°\mathrm{C})$

$P_{sat}(\mathrm{kPa}) = 2.12329 - 0.190859 \cdot T(°\mathrm{C}) + 4.35824 \times 10^{-3} \cdot T^2(°\mathrm{C}) - 3.6106 \times 10^{-5} \cdot T^3(°\mathrm{C}) + 1.08408 \times 10^{-7} \cdot T^4(°\mathrm{C})$

For higher operating temperatures (above 500°C) in a CSP system based on solar power tower technologies, nitrate molten salts, such as mixtures of sodium nitrate, sodium nitrite, and potassium nitrate, $NaNO_3$-$NaNO_2$-KNO_3, in mole fractions of 7%-49%-44%, have been widely considered for both the HTF and thermal storage. This eutectic salt mixture is also known as HITEC salt (Coastal Chemical) [8–10]; it has an upper limit temperature of 538°C and a melting point of 142°C. Freshly prepared HITEC is a white, granular solid. Liquid HITEC is pale yellow.

Similar to HITEC salt is Solar Salt, which is an eutectic nitrate salt, 60 wt % $NaNO_3$-40wt%KNO_3 in weight percentages. The melting point of the fluid is 221°C and the upper limit of its operating temperature is 550°C.

Some of the basic transport properties for HITEC and Solar Salt are given in Table A.3 [8,11].

Table A.3 Properties for HITEC salt and Solar Salt in liquid state

	Data	Correlations
HITEC salt [8] Mole: 7%$NaNO_3$–49%$NaNO_2$–44%KNO_3; Weight: 7% $NaNO_3$-40% $NaNO_2$-53%KNO_3 (MP = 142°C; unstable over 538°C)		
Density (kg/m^3)		$\rho = 2293.6 - 0.7497T$; ±2% T: (K)
Heat capacity (J/kg K)		$Cp = 5806 - 10.833T + 7.2413 \times 10^{-3}T^2$ ±5% T: (K)
Viscosity (Pa s)		$\mu = 0.4737 - 2.297 \times 10^{-3}T + 3.731 \times 10^{-6}T^2 - 2.019 \times 10^{-9}T^3$ ±16% T: (K)
Thermal conductivity (W/m K)	0.48 @ 400°C [11]	Not available
Solar Salt [11] Mole: 50%$NaNO_3$–50%KNO_3 (MP = 223°C; unstable over 550°C)		
Density (kg/m^3)		$\rho = 2263.628 - 0.636T$ T: (K)
Heat capacity (J/kg K)		$Cp = 1396.044 + 0.172T$ T: (K)
Viscosity (Pa s)		$\mu = 0.075439 - 2.77 \times 10^{-4}(T - 273) + 3.49 \times 10^{-7}(T - 273)^2 - 1.474 \times 10^{-10}(T - 273)^3$ T: (K)
Thermal conductivity (W/m K)	0.45 @ 400°C [11]	Not available

Table 3.1 Three eutectic salts from chloride salts NaCl-KCl-$ZnCl_2$

Chemical formula		NaCl	KCl	$ZnCl_2$	Theoretical melting point
Density (g/cm^3 @25°C)		2.16	1.98	2.91	
Melting Point (°C)		801	770	292	
Boiling Point (°C)		1413	1420	732	
Molar Mass (g/mol)		58.44	74.55	136.32	
#1	Molar Fraction	13.8%	41.9%	44.3%	229°C
	Mass Fraction	8.1%	31.3%	60.6%	
#2	Molar Fraction	18.6%	21.9%	59.5%	213°C
	Mass Fraction	10.0%	15.1%	74.9%	
#3	Molar Fraction	13.4%	33.7%	52.9%	204°C
	Mass Fraction	7.5%	23.9%	68.6%	

Recent research and development efforts for the new generation of HTFs for CSP technologies, led by the SunShot program of the US Department of Energy, have a new target requiring the temperatures of the HTF to reach 800°C in order to have higher energy efficiency using supercritical CO_2 thermal cycles. It has been found that eutectic chloride salts such as NaCl-KCl-$ZnCl_2$ and KCl-$MgCl_2$ can satisfy the requirement [12]. Compositions and properties for three eutectic salts, as shown in Table 3.1, with different fractions of NaCl-KCl-$ZnCl_2$ are provided in Table A.4.

Due to its lower corrosion (with nickel-based alloys) and lower cost, eutectic salt 68%KCl-32%$MgCl_2$ (molar fractions) has high potential for use as an HTF and thermal energy storage material, even with its relatively higher melting point. Some properties for this salt are collected in Table A.5.

Liquid metals and eutectic carbonate molten salts have also been proposed to serve as high-temperature HTFs by researchers [13–19].

3.2.1 Criterion for Selection of Fluids for Heat Transport Purposes

For selection of a candidate HTF, a general criterion is needed to evaluate the merit of a fluid, since multiple transport properties of a fluid will affect the heat transfer performance. The analysis in this section serves the purpose of evaluation of all the properties of a fluid, so that comparisons can be made when choosing a better HTF from the perspective of the fluid properties.

As illustrated in Fig. 3.1, the purpose of an HTF is to take the heat from a solar thermal concentrator and then deliver it to thermal storage systems or thermal power plants for electrical energy generation. A good heat transfer fluid should render a minimum temperature difference between the heat source T_{h-w} and fluid at an average temperature of $(T_i + T_o)/2$, as well as a minimum temperature difference between the fluid (at an average

Table A.4 Properties for eutectic chloride salts [12]

	Data	Correlations
#1 $NaCl\text{-}KCl\text{-}ZnCl_2$ Mole: 13.8%-41.9%-44.3%		
Density (kg/m^3)		$\rho = 2541.74 - 0.53018137763T$ ± 13.51 Unit: T(K)
Heat capacity (J/kg K)	917.0 ± 65.0	
Viscosity (Pa s)		$\mu = 152.3679\exp(-T/56.0314) + 0.05994\exp(-T/235.78682) + 2.97 \times 10^{-3}$ $\pm 1.31 \times 10^{-5}$ Unit: T(K)
Thermal conductivity (W/m K)		$k = 0.4372 - 1.2300724988 \times 10^{-4}T$ ± 0.0089 Unit: T(K)
#2 $NaCl\text{-}KCl\text{-}ZnCl_2$ Mole: 18.6%-21.9%-59.5%		
Density (kg/m^3)		$\rho = 2581.09 - 0.43205969697T$ ± 6.13 Unit: T(K)
Heat capacity (J/kg K)	913.0 ± 56.0	
Viscosity (Pa s)		$\mu = 131.0731\exp(-T/62.36328) + 4.46 \times 10^{-3}$ $\pm 2.04 \times 10^{-5}$ Unit: T(K)
Thermal conductivity (W/m K)		$k = 0.3895 - 8.1685567308 \times 10^{-5}T$ ± 0.0070 Unit: T(K)
#3 $NaCl\text{-}KCl\text{-}ZnCl_2$ Mole: 13.4%-33.7%-52.9%		
Density (kg/m^3)		$\rho = 2878.32 - 0.92630377059T$ ± 20.25 Unit: T(K)
Heat capacity (J/kg K)	900.0 ± 59.0	
Viscosity (Pa s)		$\mu = 0.12055\exp(-T/204.70939) + 497613.0848\exp(-T/29.9169) + 3.41 \times 10^{-3} \pm 2.68 \times 10^{-5}$ Unit: T (K)
Thermal conductivity (W/m K)		$k = 0.5145 - 2.3308636401 \times 10^{-4}T$ ± 0.0099 Unit: T(K)

Table A.5 Properties for eutectic chloride salts [11]

KCl-$MgCl_2$ Mole: 67%-33% close to 68%-32%		
	Data	***Correlations***
Density (kg/m^3)		$\rho = 2363.84 - 0.474T$ Unit: T (K)
Heat capacity (J/kg K)	1150 Const.	No correlation available.
Viscosity (Pa s)		$\mu = 1.46 \times 10^{-4} \exp(2230/T)$ Unit: T (K), from 873–1073 K
Thermal conductivity (W/m K)	0.4 @ 894 K	$k = 0.2469 + 5.025 \times 10^{-4} T$ Unit: T (K) In a temperature range of 730–760 K

temperature of $(T_i + T_o)/2$ and the destination heat reservoir, T_{l-w}, to which heat is delivered. Here, the destination heat reservoir is the heat source for the thermal power cycle and therefore the temperature T_{l-w} is a deterministic parameter of the thermal efficiency of the power plant.

To summarize, the requirements for the properties of an HTF used in a concentrated solar thermal power plant are multifold. First and foremost, the fluid should be able to remain as a liquid at a wide range of temperatures while having low vapor pressures to meet safety requirements of pipes and containers. Second, the fluid must have favorable properties, including density, thermal conductivity, heat capacity, and viscosity, so that overall the exergy (or useful heat) of the received thermal energy has minimal loss or degradation during the flow and heat transfer processes in the heat transmit system. Third, the fluid must have low corrosion in the pipes and containers of the heat transfer system. Last, but not least, the cost of the materials for the HTF should be sufficiently low to be acceptable by industrial customers.

When only the thermal and transport properties are of concern, the criterion for selection of fluids should be based on the principle of minimum entropy production in the heat transfer and fluid flow when accomplishing the thermal energy transport [20], since it is hard to make a fair comparison between two heat transfer fluids by simply comparing individual transport properties. For example, one fluid may have higher thermal conductivity and a higher viscosity than another fluid. While the high thermal conductivity contributes to better heat transfer, the high viscosity will contribute to larger pressure loss during the flow in the process of transporting fluid. It is therefore necessary to have one criterion that can encompass all the transport properties of a heat transfer fluid, so that comparison of benefits or advantages of HTFs can be made possible.

The entropy production comes from the two heat transfer processes as well as the friction loss in the fluid circulation, as shown in Fig. 3.1. In the following section, theoretical analysis is presented.

3.2.2 Entropy Production in the Thermal Transport Process in a CSP System

Due to the high temperature demand for the heat source in a power plant, one design point is to prescribe the desired fluid temperatures at inlet and outlet of the solar collection section as T_i and T_o, respectively. This gives an average temperature for the HTF of $(T_i + T_o)/2$ in the heat receiving solar collector. Correspondingly, after obtaining a temperature of T_o from the solar collector, the HTF flows into the heat exchanger to deliver heat to the working fluid of the power cycle, which makes the HTF temperature return to T_i. For a given heat transfer rate, $\dot{Q}$, and the temperatures of T_i and T_o, a good HTF should be able to keep the wall temperature T_{h-w} in the solar collector not too high, and also the wall temperature T_{l-w} in the heat delivery heat exchanger should not be too low. The pumping power to achieve this goal of thermal transport should also be low.

The dimensions of pipes and heat exchangers in the system illustrated in Fig. 3.1 have been prescribed based on the properties of a known HTF and the demanded heat transfer rate $\dot{Q}$, which is the total thermal energy supply per unit of time for a CSP plant. The entropy production rates in the system can thus be obtained for an HTF, and comparison of the entropy production between different HTFs is possible. The one that introduces the minimum entropy production in the system will be selected as the best HTF.

From the second law of thermodynamics, the steady-state entropy production rate in the system (a control volume from location 1 to location 2, including the pipes and heat exchangers that the HTF fills) should be expressed as:

$$s_2\dot{m} = s_1\dot{m} + \int\limits_{collector} \frac{\delta\dot{Q}_{in}}{T_{h-w}} - \int\limits_{exchanger} \frac{\delta\dot{Q}_{out}}{T_{l-w}} - \dot{m}\frac{v}{T_m}\Delta P + \dot{m}s_{gen} \tag{3.5}$$

where the system inlet is at location 1 and the outlet is at location 2; T_{h-w} is the wall temperature of the heat source (concentrated solar energy) that is added to the system, and T_{l-w} is the wall temperature of the heat exchanger where the HTF gives out heat.

In order to conveniently observe how fluid properties affect the entropy production in the system, the heat transfer in the solar collector and the heat exchanger in Fig. 3.1 is assumed to have uniform heat flux across the boundary walls. In fact, for a solar heat collector and a countercurrent heat exchanger, this assumption is quite reasonable. Therefore, the integral terms $\int_{collector} (1/T_{h-w})\delta\dot{Q}_{in}$ and $\int_{exchanger} (1/T_{l-w})\delta\dot{Q}_{out}$ in Eq. (3.5) can be simplified to ratios of the total heat rate and the average temperatures of the walls, which gives $\dot{Q}_{in}/T_h$ and $\dot{Q}_{out}/T_l$. Here T_h and T_l are the average values respectively from local T_{h-w} and T_{l-w}. A numerical test can easily demonstrate the accuracy of approximating the term $\int_{collector} (1/T_{h-w})\delta\dot{Q}_{in}$ by $\dot{Q}_{in}/T_h$ for constant heat flux and linear variation of temperature T_{h-w}. The same is true for $\int_{collector} (1/T_{h-w})\delta\dot{Q}_{in}$ to be approximated by $\dot{Q}_{in}/T_h$.

Since the HTF circulates in the system, we can choose points 1 and 2 to be the same point, meaning the fluid circulates from a starting point and back to an ending point which is the same as the starting point. Therefore, there must be $s_2\dot{m} = s_1\dot{m}$, which results in the following expression for the system entropy production rate:

$$\dot{m}s_{gen} = \dot{S}_{gen} = \frac{\dot{Q}_{out}}{T_l} - \frac{\dot{Q}_{in}}{T_h} + \dot{m}\frac{v}{T_m}\Delta P = \frac{\dot{Q}_{out}}{T_l} - \frac{\dot{Q}_{in}}{T_h} + \frac{\dot{V}}{T_m}\Delta P \quad (3.6)$$

where $\dot{S}_{gen}$ is the system entropy production in a steady-state operation, and $\dot{V}$ is the volume flow rate of the HTF; $\dot{Q}_{in}$ and $\dot{Q}_{out}$ are the absolute values of the heat rates added to and delivered out of the system, respectively. It is clear to see from Eq. (3.6) that the entropy production is due to the two heat transfer processes and the pressure loss in the system that is overcome by the pump. Ideally, the heat added into the system and that removed from the system have the same absolute value. For convenience, we have $\dot{Q}_{out} = \dot{Q}_{in} = \dot{Q}$ in the analyses hereafter.

The fluid temperature at the inlet of the solar collection field is defined as T_i and at the exit of the solar collection field as T_o. When considering the heat removal in the heat exchanger to deliver the heat to the fluid in the thermal power cycle, it is easy to understand that the temperature of the HTF at the inlet and exit of this heat exchanger is T_o and T_i, respectively, which is just the opposite of that in the solar collection field. The average temperatures of the HTF in the solar collection field and the heat exchanger are therefore the same, which is $T_m = 0.5(T_o + T_i)$. The average temperature of the HTF in circulation should also be $T_m = 0.5(T_o + T_i)$. The temperature differences that drive the heat exchange in the solar collection field and in the heat exchanger are $T_h - 0.5(T_o + T_i)$ and $0.5(T_o + T_i) - T_l$, respectively.

Considering the heat transfer processes in the solar collection field and in the heat exchanger, as seen in Fig. 3.1, there are:

$$\dot{Q} = A_h h_h [T_h - 0.5(T_o + T_i)] \tag{3.7}$$

$$\dot{Q} = A_l h_l [0.5(T_o + T_i) - T_l] \tag{3.8}$$

where A_h and A_l are heat transfer surface areas; h_h and h_l are heat transfer coefficients in the solar collector and heat exchanger, respectively. From these two equations, we can find that

$$T_h = \frac{\dot{Q}}{A_h h_h} + 0.5(T_o + T_i) \tag{3.9}$$

$$T_l = 0.5(T_o + T_i) - \frac{\dot{Q}}{A_l h_l} \tag{3.10}$$

Substituting Eqs. (3.9), (3.10) into Eq. (3.6), the system entropy production rate is obtained:

$$\dot{S}_{gen} = \frac{\dot{Q}}{0.5(T_o + T_i) - \dot{Q}/(A_l h_l)} - \frac{\dot{Q}}{0.5(T_o + T_i) + \dot{Q}/(A_h h_h)} + \frac{\dot{V}}{0.5(T_o + T_i)} \Delta P \tag{3.11}$$

For the purpose of evaluating the transport properties of an HTF, this analysis needs to give the demanded heat rate $\dot{Q}$ and the desired temperatures of T_i and T_o. With the given transport properties of the HTF, the flow rates of the HTF can be decided:

$$\dot{Q} = \dot{m} Cp(T_o - T_i) \tag{3.12}$$

$$\dot{m} = \dot{V} \rho \tag{3.13}$$

$$\dot{V} = \frac{\dot{Q}}{\rho Cp(T_o - T_i)} \tag{3.14}$$

Substituting Eq. (3.14) into Eq. (3.11), the entropy production rate is further expressed as

$$\dot{S}_{gen} = \frac{\dot{Q}}{0.5(T_o + T_i) - \dot{Q}/(A_l h_l)} - \frac{\dot{Q}}{0.5(T_o + T_i) + \dot{Q}/(A_h h_h)} + \frac{\dot{Q}}{0.5 \rho Cp(T_o + T_i)(T_o - T_i)} \Delta P \tag{3.15}$$

where the first two terms are due to the heat transfer processes, which is named $\dot{S}_{gen-ht}$, and the last term is due to pressure loss when the HTF circulates in the system (including in heat exchangers and in circulation pipes), which may be defined as $\dot{S}_{gen-fl}$.

From Eq. (3.15), it is also understandable that large values of h_l and h_h will result in less entropy production, while the pressure loss ΔP will contribute to higher entropy production. Furthermore, it is also easy to understand that if the heat transfer is improved (either due to heat transfer surface enhancement or better properties of the HTF), the entire heat transfer task can be achieved with the temperature T_l closer to T_h, which is beneficial to the thermal energy efficiency in the thermal power plant. The pressure loss in the entire system includes the losses from two heat exchangers and from fluid circulation pipes, which may be defined as ΔP_{HE-h}, ΔP_{HE-l}, and ΔP_c, respectively. With the volume flow rate of the HTF available from Eq. (3.14), the pressure losses in the system can be conveniently obtained.

It is known that the heat transfer coefficients, h_h and h_l, and the pressure loss, ΔP, in the circulation of fluid flow can be expressed in terms of the fluid properties, the prescribed temperatures, as well as the demand of the heat transfer rate, which determines the flow rate and velocities of the fluid in given devices.

It is noted that Eq. (3.15) is the basic expression for the entropy generation. There are various equations available for one to find the heat transfer coefficients h_h and h_l in various types of heat exchangers, and the pressure loss ΔP in the fluid flow in the circulation [21–26]. As an example, the following analysis picked up some equations for h_h, h_l, and ΔP to demonstrate how fluid properties affect the entropy generation.

For the heat transfer of turbulent flow in pipes with Reynolds number (Re) close to 1.0×10^4 or higher, $0.7 \leq \mathrm{Pr} \leq 160$, and length-diameter ratio $L/d \geq 10$, we use the famous Dittus-Boelter equation [22] to obtain heat transfer coefficients as follows:

$$
\begin{aligned}
Nu_h &= \frac{h_h d_h}{k} = 0.023 \mathrm{Re}_h^{0.8} \mathrm{Pr}^{0.4} \text{ (when fluid is heated) or} \\
Nu_l &= \frac{h_l d_h}{k} = 0.023 \mathrm{Re}_l^{0.8} \mathrm{Pr}^{0.3} \text{ (when fluid is cooled)}
\end{aligned}
\tag{3.16}
$$

Assume the heat transfer design analysis already found the number (n) of tubes (with diameter d) for a heat exchanger. Therefore, the flow velocity in the tubes of a heat exchanger is obtained as:

$$u = \frac{\dot{V}}{n(0.25\pi d^2)} = \frac{\dot{Q}}{\rho Cp(T_o - T_i)n(0.25\pi d^2)} = \frac{\dot{Q}}{\rho Cp(T_o - T_i)Nd^2} \tag{3.17}$$

where we define $N = 0.25n\pi$. Using Eq. (3.14), the Reynolds number for the flow in the heat transfer tubes is:

$$\mathrm{Re} = \frac{\dot{Q}}{\rho Cp(T_o - T_i)Nd^2}\frac{\rho d}{\mu} = \frac{\dot{Q}}{Cp(T_o - T_i)N \cdot d \cdot \mu} \tag{3.18}$$

Bringing the expression for Reynolds number into Eq. (3.16), the heat transfer coefficients are obtained as follows:

$$\begin{aligned} h_h &= \frac{k}{d_h}0.023\left[\frac{\dot{Q}}{\rho Cp(T_o - T_i)N_h {d_h}^2}\frac{\rho d_h}{\mu}\right]^{0.8}\left(\frac{\mu Cp}{k}\right)^{0.4} \\ &= 0.023\left[\frac{\dot{Q}}{N_h(T_o - T_i)}\right]^{0.8}\frac{1}{{d_h}^{1.8}}\frac{k^{0.6}}{Cp^{0.4}\mu^{0.4}} \\ h_l &= \frac{k}{d_l}0.023\left[\frac{\dot{Q}}{\rho Cp(T_o - T_i)N_l {d_l}^2}\frac{\rho d_l}{\mu}\right]^{0.8}\left(\frac{\mu Cp}{k}\right)^{0.3} \\ &= 0.023\left[\frac{\dot{Q}}{N_l(T_o - T_i)}\right]^{0.8}\frac{1}{{d_l}^{1.8}}\frac{k^{0.7}}{Cp^{0.5}\mu^{0.5}} \end{aligned} \tag{3.19}$$

Substituting Eq. (3.19) into Eq. (3.15), the entropy generation due to heat transfer in the system is

$$\begin{aligned} \dot{S}_{gen-ht}^{Turbulent} = & \frac{\dot{Q}}{0.5(T_o + T_i) - \frac{\dot{Q}}{4N_l^{0.2}L_l 0.023}\frac{1}{}\left[\frac{(T_o - T_i)}{\dot{Q}}\right]^{0.8}\frac{d_l^{0.8}Cp^{0.5}\mu^{0.5}}{k^{0.7}}} \\ & - \frac{\dot{Q}}{0.5(T_o + T_i) + \frac{\dot{Q}}{4N_h^{0.2}L_h 0.023}\left[\frac{(T_o - T_i)}{\dot{Q}}\right]^{0.8}\frac{d_h^{0.8}Cp^{0.4}\mu^{0.4}}{k^{0.6}}} \end{aligned} \tag{3.20}$$

The equations here explicitly show us the impact of fluid properties on the entropy production in heat transfer processes.

It is critical to point out that the total entropy production analysis in this chapter is very useful in the evaluation of a thermal energy transport system in a concentrated solar power plant. The thermodynamic tool used here provides fundamental methodology and does not restrict the analysis to any specific heat transfer fluids or designs for the heat exchange devices. Therefore, this type of analysis allows engineers to conduct a fair comparison between different ideas of heat transport system designs, using gases, liquid, or even solid particles for the heat transfer media.

3.2.3 Entropy Production Rate Due to Pressure Losses

The following equation [22] is chosen to calculate the pressure loss due to turbulent flow in the heat transfer tubes in heat exchangers:

$$\Delta P_{HE} = \frac{f \cdot \rho u^2 L}{2d} \tag{3.21}$$

where f is the Darcy friction factor. If the flow Reynolds number is $2300 \leq \mathrm{Re} \leq 2 \times 10^4$, using the Blasius equation ($f = 4 \cdot 0.079\mathrm{Re}^{-0.25}$) for the Darcy friction factor, the pressure drop is:

$$\Delta P_{HE} = 0.079 \frac{2 \cdot \rho u^2 L}{d} \left(\frac{u d \rho}{\mu}\right)^{-0.25} = 0.158 \frac{\rho^{0.75} u^{1.75} L \mu^{0.25}}{d^{1.25}} \tag{3.22}$$

where L is the length of the tube or pipe subject to study. Substituting velocity u (from Eq. (3.17)) into Eq. (3.22):

$$\begin{aligned} \Delta P_{HE} &= 0.158 \frac{\rho^{0.75} L \mu^{0.25}}{d^{1.25}} \left[\frac{\dot{Q}}{\rho Cp (T_o - T_i) N d^2}\right]^{1.75} \\ &= 0.158 \frac{L \mu^{0.25}}{\rho Cp^{1.75} d^{4.75}} \left[\frac{\dot{Q}}{(T_o - T_i) N}\right]^{1.75} \end{aligned} \tag{3.23}$$

For the pressure loss in the fluid circulation pipes (with a total length of L_c and pipe diameter of d_c), the same expression of pressure loss is:

$$\begin{aligned} \Delta P_c &= 0.158 \frac{\rho^{0.75} L_c \mu^{0.25}}{d_c^{1.25}} \left[\frac{\dot{Q}}{\rho Cp (T_o - T_i) N_c d_c^2}\right]^{1.75} \\ &= 0.158 \frac{L_c \mu^{0.25}}{\rho Cp^{1.75} d_c^{4.75}} \left[\frac{\dot{Q}}{(T_o - T_i) N_c}\right]^{1.75} \end{aligned} \tag{3.24}$$

where N_c is defined as $N_c = 0.25 n_c \pi$, and n_c is the number of pipes that are used to circulate the fluid.

The total system pressure drop includes three parts, from the fluid circulation pipes and from the heat exchangers (one has heat added to the HTF, and the other has heat removed from the HTF). Therefore, the total pressure drop in the heat exchangers and in the circulation pipe is:

$$\Delta P = \Delta P_{HE-h} + \Delta P_{HE-l} + \Delta P_c \tag{3.25}$$

Substituting pressure losses into the third term in Eq. (3.11), the entropy production rate due to pressure loss is

$$\dot{S}^{T-1}_{gen-fl} = \frac{\dot{Q}}{0.5(T_o + T_i)(T_o - T_i)} \frac{0.158\mu^{0.25}}{\rho^2 Cp^{2.75}} \left[\frac{\dot{Q}}{(T_o - T_i)}\right]^{1.75} \left(\frac{L_h}{d_h^{4.75} N_h^{1.75}} + \frac{L_l}{d_l^{4.75} N_l^{1.75}} + \frac{L_c}{d_c^{4.75} N_c^{1.75}}\right) \tag{3.26}$$

where $\dot{S}^{T-1}_{gen-fl}$ designates the first case of turbulent flow, where $2300 \leq \mathrm{Re} \leq 2 \times 10^4$.

Under the condition of $\mathrm{Re} \geq 2 \times 10^4$, the equation for pressure loss of turbulent flow in the heat transfer tubes of heat exchangers is:

$$\Delta P_{HE} = 0.184 \frac{\rho u^2 L}{2d} \left(\frac{ud\rho}{\mu}\right)^{-0.2} = 0.092 \frac{\rho^{0.8} u^{1.8} L \mu^{0.2}}{d^{1.2}} \tag{3.27}$$

Bringing the expression for velocity given in Eq. (3.17) into the preceding expression:

$$\Delta P_{HE} = 0.092 \frac{\rho^{0.8} L \mu^{0.2}}{d^{1.2}} \left[\frac{\dot{Q}}{\rho Cp (T_o - T_i) N d^2}\right]^{1.8} = 0.092 \frac{L \mu^{0.2}}{\rho Cp^{1.8} d^{4.8}} \left[\frac{\dot{Q}}{(T_o - T_i) N}\right]^{1.8} \tag{3.28}$$

Similarly, the pressure drop in a circulation pipe with $\mathrm{Re} \geq 2 \times 10^4$ is:

$$\Delta P_c = 0.092 \frac{\rho^{0.8} L_c \mu^{0.2}}{d_c^{1.2}} \left[\frac{\dot{Q}}{\rho Cp (T_o - T_i) N_c d_c^2}\right]^{1.8} = 0.092 \frac{L_c \mu^{0.2}}{\rho Cp^{1.8} d_c^{4.8}} \left[\frac{\dot{Q}}{(T_o - T_i) N_c}\right]^{1.8} \tag{3.29}$$

In cases where the Reynolds numbers of the flow in the heat exchangers and in the HTF circulation pipes are all larger than 2×10^4, the entropy production rate due to all pressure losses is:

$$\dot{S}^{T-2}_{gen-fl} = \frac{\dot{Q}}{0.5(T_o + T_i)(T_o - T_i)} \frac{0.092\mu^{0.2}}{\rho^2 Cp^{2.8}} \left[\frac{\dot{Q}}{(T_o - T_i)}\right]^{1.8} \left(\frac{L_h}{d_h^{4.8} N_h^{1.8}} + \frac{L_l}{d_l^{4.8} N_l^{1.8}} + \frac{L_c}{d_c^{4.8} N_c^{1.8}}\right) \tag{3.30}$$

where $\dot{S}^{T-2}_{gen-fl}$ designates the second case of turbulent flow when $\mathrm{Re} \geq 2 \times 10^4$.

The preceding analysis provides equations for the pressure loss at different Reynolds numbers. When the flow Reynolds numbers in two heat exchangers and the circulation loop are different, one can always choose corresponding equations for heat transfer coefficients and pressure loss and use the basic equation (Eq. 3.15) for the entropy generation.

3.2.4 Discussion and Examples of Evaluation of several Heat Transfer Fluids

In all the preceding expressions of entropy production, the given parameters are the system heat transfer/transport rate $\dot{Q}$, the desired working temperatures of fluid T_i and T_o, and dimensions of the devices including d_h, d_l, d_c, L_h, L_l, L_c, n_h, n_l, and n_c. In fact, the dimensions (L_h, L_l, n_h, n_l) of the devices must be obtained through design of the heat exchangers, which is based on constraints of given diameters of the heat transfer tubes and the tolerable maximum temperature of the tube material against the desired temperatures of fluid T_i and T_o.

For heat transfer of turbulent flow (which includes most practical applications), it is interesting to examine Eq. (3.20) and note that the clusters of properties $d_l^{0.8}Cp^{0.5}\mu^{0.5}/k^{0.7}$ and $d_h^{0.8}Cp^{0.4}\mu^{0.4}/k^{0.6}$ actually influence the entropy production in the heat transfer processes. A smaller value of the cluster will result in less entropy production in the system.

For the entropy production due to pressure loss in the heat exchangers and in the circulation pipe, the property clusters are $\mu^{0.25}/(\rho^2Cp^{2.75}d^{4.75})$ and $\mu^{0.2}/(\rho^2Cp^{2.8}d^{4.8})$ for turbulent flow with Reynolds number less than 2×10^4, and larger than 2×10^4, respectively. The smaller values of these clusters will result in less entropy production in the system, which is favorable. The clusters of HTF properties contributing to larger entropy generation due to heat transfer and pressure drop are summarized in Table 3.2.

It is important to note that a large viscosity always contribute to a large entropy production either from heat transfer or from pressure loss. A high thermal conductivity contributes to less entropy production in the heat transfer process. However, the effect of fluid heat capacity on the entropy

Table 3.2 Clusters of properties contributing to entropy production in the system

Heat transfer		Pressure drop	
Fluid is heated	**Fluid is cooled**		
$\mathrm{Re}\geq1\times10^4$: $d_h^{0.8}Cp^{0.4}\mu^{0.4}/k^{0.6}$	$\mathrm{Re}\geq1\times10^4$: $d_l^{0.8}Cp^{0.5}\mu^{0.5}/k^{0.7}$	$\mathrm{Re}\leq2\times10^4$: $\mu^{0.25}/(\rho^2Cp^{2.75}d^{4.75})$	$\mathrm{Re}\geq2\times10^4$: $\mu^{0.2}/(\rho^2Cp^{2.8}d^{4.8})$

production in heat transfer processes is interesting to examine. From Eq. (3.20), it seems that a large fluid Cp contributes to higher entropy production. On the other hand, because of large heat capacity, the heat transfer task can be achieved with less mass flow, which means that smaller flow tubes can be used. This is seen in the cluster of properties, $d_l^{0.8}Cp^{0.5}\mu^{0.5}/k^{0.7}$, where with large Cp the tube diameter d_l may be reduced.

The following is an example that demonstrates the system entropy production for transferring an amount of 50 MW thermal energy. The working temperatures, fluids and their properties, as well as the obtained system entropy production for four different working temperatures using the previously presented methods, are given in Tables 3.3–3.6. To summarize and

Table 3.3 Entropy production rate for transporting 50 MW thermal energy at $T_m = 200°C$ using three different HTFs

	XCELTHERM 600 at 200°C	Therminol VP-1 at 200°C	Lead-bismuth eutectic (LBE)
ρ (kg/m^3)	743.2	913	Equation in Table 3.7
μ (m Pa s)	0.623	0.395	Equation in Table 3.7
k (W/m K)	0.1216	0.114	Equation in Table 3.7
Cp (kJ/kg K)	2.61	2.048	Equation in Table 3.7
T_h (K)	573	552	(484.9)
T_l (K)	373	399	(461.0)
$\dot{S}_{gen}$ (J/(K s))	46,793	34,872	(5432.4)
$\eta = 1 - 300/T_l$ (K)	0.196	0.247	(0.34929)

$T_i = 423$ K, $T_o = 523$ K, $d_l = d_h = 0.02$ m, $n_l = n_h = 820$, $L_h = 7.7$ m, $L_l = 7.7$ m, $d_c = 0.5$ m, $n_c = 4$, $L_c = 200$ m.

Table 3.4 Entropy production rate for transporting 50 MW thermal energy at $T_m = 340°C$ using three different HTFs

	Therminol VP-1 at 340°C	KNO_3-$NaNO_2$-$NaNO_3$ (7%-49%-44% mole) at 340°C	Lead-bismuth eutectic (LBE)
ρ (kg/m^3)	773	1833.921	Equation in Table 3.7
μ (m Pa s)	0.185	2.56394	Equation in Table 3.7
k (W/m K)	0.089	0.51	Equation in Table 3.7
Cp (kJ/kg K)	2.425	1.89	Equation in Table 3.7
T_h (K)	723	713	(639.2)
T_l (K)	523	526	(586.8)
$\dot{S}_{gen}$ (J/(K s))	26,448	24,956	(7023.2)
$\eta = 1 - 300/T_l$ (K)	0.426	0.429	(0.48877)

$T_i = 563$ K, $T_o = 663$ K, $d_l = d_h = 0.02$ m, $n_l = n_h = 820$, $L_h = 5.1$ m, $L_l = 5.1$ m, $d_c = 0.5$ m, $n_c = 4$, $L_c = 200$ m.

Table 3.5 Entropy production rate for transporting 50 MW thermal energy at $T_m = 450°C$ using three different HTFs

	KNO_3-$NaNO_2$-$NaNO_3$ (7%-49%-44% mole) at 450°C	NaCl-KCl-$ZnCl_2$ at 450°C	Lead-bismuth eutectic (LBE)
ρ (kg/m^3)	1751.454	2150	Equation in Table 3.7
μ (m Pa s)	1.375	7.0	Equation in Table 3.7
k (W/m K)	0.51	0.32	Equation in Table 3.7
Cp (kJ/kg K)	1.7596	0.81	Equation in Table 3.7
T_h (K)	793	853	(740.2)
T_l (K)	623	511	(705.8)
$\dot{S}_{gen}$ (J/(K s))	17,206	39,287	(3327.4)
$\eta = 1 - 300/T_l$ (K)	0.518	0.413	(0.57497)

$T_i = 673$ K, $T_o = 773$ K, $d_l = d_h = 0.02$ m, $n_l = n_h = 820$, $L_h = 5.5$ m, $L_l = 5.5$ m, $d_c = 0.5$ m, $n_c = 4$, $L_c = 200$ m.

Table 3.6 Entropy production rate for transporting 50 MW thermal energy at $T_m = 700°C$ using three different HTFs

	NaCl-KCl-$ZnCl_2$ 700°C	NaCl-KCl-$ZnCl_2$ 700°C (Target)	Lead-bismuth eutectic (LBE)
ρ (kg/m^3)	1950	1950	Equation in Table 3.7
μ (m Pa s)	4.0	4.0	Equation in Table 3.7
k (W/m K)	0.31	0.51 (target)	Equation in Table 3.7
Cp (kJ/kg K)	0.90	1.50 (target)	Equation in Table 3.7
T_h (K)	1083	1074	(989.3)
T_l (K)	800	815.5	(956.7)
$\dot{S}_{gen}$ (J/(K s))	16,355	14,745	(1751.6)
$\eta = 1 - 300/T_l$ (K)	0.625	0.632	(0.686)

$T_i = 923$ K, $T_o = 1023$ K, $d_l = d_h = 0.02$ m, $n_l = n_h = 820$, $L_h = 5.5$ m, $L_l = 5.5$ m, $d_c = 0.5$ m, $n_c = 4$, $L_c = 200$ m.

compare the figure-of-merit of different HTFs, Fig. 3.2 shows the entropy production rates of three different HTFs at each operating temperature, T_m.

The preceding example provides a standard process and equations for evaluating the figure-of-merit (FOM) of HTFs that transfer heat from a concentrated solar collector and deliver to a thermal power plant. Entropy production of the entire system is compared for different HTFs.

To be noted is that, through analysis, the roles and contributions of all the transport properties to the minimized system entropy generation could be found, and an FOM including all the transport properties was able to be

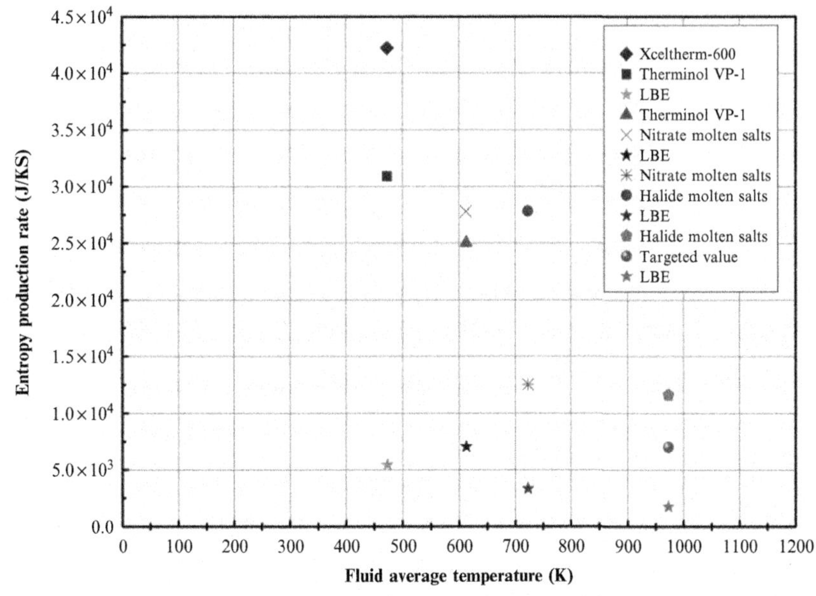

Fig. 3.2 The entropy production rate of the system using different HTFs at different operating temperatures for the task of transferring 50 MW thermal energy.

defined. Since the entropy production is related to the exergy destruction (defined as $T_0\dot{S}_{gen}$) or the reduction of useful work, this analysis can eventually be associated with cost analysis for the system.

It is also important to note that the analyses in the preceding work only evaluate the transport properties of HTFs. Other issues, such as material cost,

Table 3.7 Summary of correlations for properties of molten LBE ($P \sim 0.1$ MPa)

Properties	Unit	Correlations T(K)	Temperature range (K)	Uncertainty
ρ	kg m^{-3}	$\rho = 11{,}096 - 1.3236\ T$	403–1300	0.8%
Cp	J/kg K	$C_p = 159 - 2.72 \times 10^{-2}\ T + 7.12 \times 10^{-6}\ T^2$	430–605 (Extrapolation applicable in 430–1900 K)	7%
μ	Pa s	$\mu = 4.94 \times 10^{-4} \times \exp(754.1/T)$	400–1100	5%
k	W/m K	$k = 3.61 + 1.517 \times 10^{-2}\ T - 1.741 \times 10^{-6}\ T^2$	403–1100	5%

Constant heat flux: $Nu = 6.3 + 0.0167\mathrm{Re}^{0.85}\mathrm{Pr}^{0.93}$ at $0.004 \leq \mathrm{Pr} \leq 0.1$ and $10^4 \leq \mathrm{Re} \leq 10^6$.

corrosion of pipes and containers by the HTF, and the needed head (pumping head) to lift the fluid to a high tower, can also significantly affect the selection of the HTF. Therefore, the current FOM still cannot serve as the one ultimate criterion for judging an HTF. Nevertheless, this work still offers a very important opportunity leading to the development of a thorough FOM, which should be the cost of electrical energy of a CSP plant with all factors considered.

3.3 SOLID MATERIALS FOR SENSIBLE THERMAL STORAGE

Solid materials, such as concrete, sand, rock, brick, soil, graphite, silicon carbide, taconite, cast iron, and even waste metal chips, have been considered or applied for thermal energy storage purposes. Some of these materials (such as rocks) can be used to form a packed bed thermal storage system with an HTF passing through the porous zone to store or withdraw thermal energy, as illustrated in Fig. 2.5a–c of Chapter 2. Some other materials may form an integrated entity (such as concrete) with pipes embedded, and heat transfer fluids flowing in the pipes to store or withdraw the thermal energy from the materials, as shown in Fig. 2.5d of Chapter 2.

The key parameter of these materials for thermal energy storage is the product of density and heat capacity ($\rho \cdot C$), as well as thermal conductivity. While the former determines the volumetric energy storage capacity, the latter is important to the thermal energy storage efficiency. High thermal conductivity will result in less exergy loss due to the heat transfer processes in the heat charging and discharging processes.

In Table 3.8 some of the properties of thermal storage materials are listed for reference [27].

3.4 SOLID PARTICLES AND HEAT TRANSFER FLUIDS MIXTURE FOR THERMAL STORAGE

Solid particles, such as sands, filled with thermal conductive liquid have also been proposed as thermal storage materials. Pipes embedded in the material have fluid flowing through to deliver or withdraw thermal energy.

A mixture of particles and a thermal conductive fluid will have higher thermal storage capacity because the liquid fills in the voids in the sands. The thermal conductivity of the mixture can also be improved significantly due to better thermal conductivity of the fluid than the sands and the air in

Table 3.8 Properties of solid materials for thermal energy storage

Medium	Melting (°C) or crumbles	ρ(kg/m^3)	C(kJ/kg · °C)	$\rho \cdot C$(kJ/m^3 · °C)	k(W/m · °C)
Aluminum	660	2700	0.92	2484.0	250
Brick (common)	1800	1920	1.0	1920.0	1.04
Fireclay	1800	2100–2600	1.0	2100–2600	1–1.5
Soil (dry)	1650	1200–1600	1.26	1512–2016	1.5
Granite	1215	2400	0.79	1896	1.7–4.0
Sand (dry)	1500	1555	0.8	1244	0.15–0.25
Sandstone	1300	2000–2600	0.92	1840–2392	2.4
Rocks	1800	2480	0.84	2086.6	2–7
Concrete	1000 (Crumbles)	2240–2400	0.75	1680–1800	1.7
Graphite	3500	2300–2700	0.71	1633–1917	85
Silicon carbide	2730	3210	0.75	2407.5	3.6
Taconite	1538	3200	0.8	2560	1.0–2.0
Cast iron	1150	7200	0.54	3888	42–55

the voids. The thermal conductivity of a sand–fluid mixture can be predicted from the following Hamilton-Crosser equation [28]:

$$k_{mix} == \frac{k_f\left[k_s + 2k_f - 2\phi\left(k_f - k_s\right)\right]}{k_s + 2k_f + \phi\left(k_f - k_s\right)} \quad (3.31)$$

or another similar equation [29] in the form of

$$k_{mix} = k_f \frac{1 + 2\beta\phi + \left(2\beta^3 - 0.1\beta\right)\phi^2 + \phi^3 0.05 \exp\left(4.5\beta\right)}{1 - \beta\phi} \quad (3.31)'$$

where $\phi = 1 - \varepsilon$, and ε is the porosity; k_{mix}, k_f, and k_s are thermal conductivities of the mixture, fluid, and generic solid particles, respectively; and $\beta = \left(k_s - k_f\right)/\left(k_s + 2k_f\right)$.

The mixing rule equations for density ρ_{mix} and heat capacity Cp_{mix} are as follows:

$$\rho_{mix} = \varepsilon \cdot \rho_f + \phi\rho_s \quad (3.32)$$

$$Cp_{mix} = \left(\varepsilon \cdot \rho_f Cp_f + \phi\rho_s Cp_s\right) / \left(\varepsilon \cdot \rho_f + \phi\rho_s\right) \quad (3.33)$$

Using these equations, the properties of sand-air and sand-oil (XCELTHERM 600) mixtures are predicted and given in Tables 3.9 and 3.10 for comparison. It is very clear that thermal conductivity, density,

Table 3.9 Properties of air, sand, and sand-air mixture

Sand name	$k(l)$ W/(m K)	$k(s)$ W/(m K)	ε	$1-\varepsilon$	k(mix) W/(m K)
Silver	0.026	1.4	0.39	0.61	0.132
Medium	0.026	1.4	0.4	0.6	0.128
Filter	0.026	1.4	0.41	0.59	0.125
Coarse	0.026	1.4	0.38	0.62	0.137
	$\rho(l)$ kg/m^3	$\rho(s)$ kg/m^3	ε	$1-\varepsilon$	ρ(mix) kg/m^3
Silver	1.067	2410	0.39	0.61	1470.516
Medium	1.067	2410	0.4	0.6	1446.427
Filter	1.067	2410	0.41	0.59	1422.337
Coarse	1.067	2410	0.38	0.62	1494.605
	$Cp(l)$ kJ/(kg K)	$Cp(s)$ kJ/(kg K)	ε	$1-\varepsilon$	Cp(mix) kJ/(kg K)
Silver	1.005	705	0.39	0.61	704.801
Medium	1.005	705	0.4	0.6	704.792
Filter	1.005	705	0.41	0.59	704.783
Coarse	1.005	705	0.38	0.62	704.809

Table 3.10 Properties of oil, sand, and sand-oil mixture

Sand Name	$k(l)$ W/(m K)	$k(s)$ W/(m K)	ε	$1-\varepsilon$	$k(\text{mix})$ W/(m K)
Sand Name	$k(l)$ W/(m K)	$k(s)$ W/(m K)	ε	$1-\varepsilon$	$k(\text{mix})$ W/(m K)
Silver	0.1347	1.4	0.39	0.61	0.482
Medium	0.1347	1.4	0.4	0.6	0.472
Filter	0.1347	1.4	0.41	0.59	0.462
Coarse	0.1347	1.4	0.38	0.62	0.493
	$\rho(l)$ kg/m^3	$\rho(s)$ kg/m^3	ε	$1-\varepsilon$	$\rho(\text{mix})$ kg/m^3
Silver	841	2410	0.39	0.61	1798.090
Medium	841	2410	0.4	0.6	1782.400
Filter	841	2410	0.41	0.59	1766.710
Coarse	841	2410	0.38	0.62	1813.780
	$Cp(l)$ kJ/(kg K)	$Cp(s)$ kJ/(kg K)	ε	$1-\varepsilon$	$Cp(\text{mix})$ kJ/(kg K)
Silver	2050	705	0.39	0.61	950.342
Medium	2050	705	0.4	0.6	958.848
Filter	2050	705	0.41	0.59	967.505
Coarse	2050	705	0.38	0.62	941.983

and heat capacity of sand-oil are all larger than those of sand-air. The thermal conductivity of sand-oil is three times that of sand-air.

The sand-oil mixture has better thermophysical properties compared to those of sand alone. As shown in Fig. 3.3, the liquid oil can fill in the gap between the embedded pipes and sands due to mismatch of thermal expansion. It has been demonstrated that sand-oil or a sand and molten salt mixture may be able to compete with concrete for thermal storage at higher temperatures to avoid issues of cracking in concrete that deteriorate the heat transfer [30].

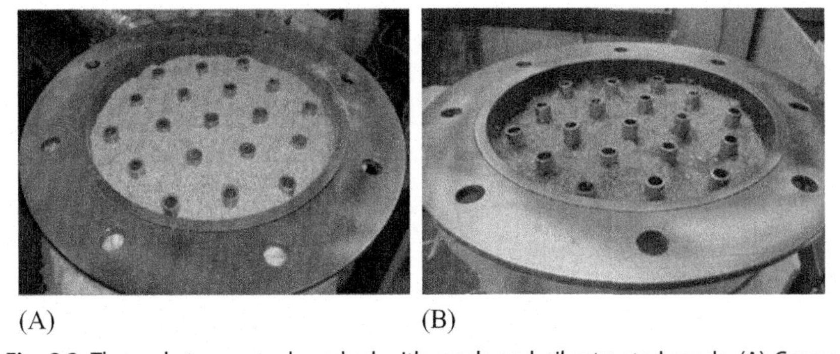

(A) (B)

Fig. 3.3 Thermal storage tank packed with sands and oil-saturated sands. (A) Coarse sand, (B) oil saturated coarse sand.

3.5 PHASE CHANGE MATERIALS FOR LATENT HEAT THERMAL STORAGE

As previously defined, materials that are able to store thermal energy during a phase change process are called PCMs. Since latent heat storage is much larger than sensible heat, PCM thermal storage has much higher energy density. A number of studies have reported various types of PCMs for thermal storage application. These materials include organic compounds, inorganic salts and metals, and eutectic salts or metals as appropriate. Fig. 3.4 lists a categorization for PCMs. Organic compound PCMs include paraffin waxes, esters, acids and alcohols; inorganic PCMs include salts, salt hydrates, eutectics of inorganic salts, and metals and their eutectics. PCMs from organic compounds generally have low melting points and can only be used for room-heating thermal storage. For high-temperature thermal storage, molten salts have been widely considered by researchers. Since molten metals and alloys are considered to be HTFs in nuclear power plants, they are viewed as possible HTFs as well as PCMs for thermal energy storage.

From the working temperature point of view, Hoshi et al. [31] categorized PCMs with melting points below 220°C as "low" temperature materials, melting temperatures up to 420°C as "medium" temperature materials, and melting points greater than 420°C as "high" temperature materials suitable for CSP thermal storage.

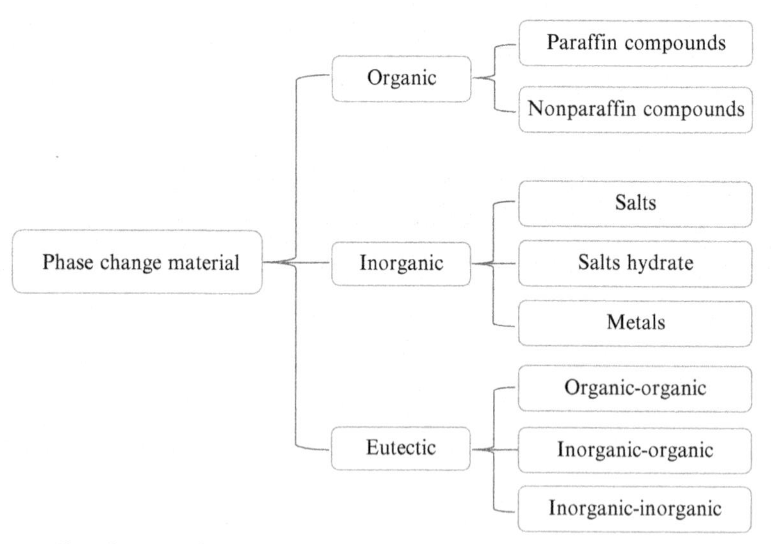

Fig. 3.4 Classification of PCMs.

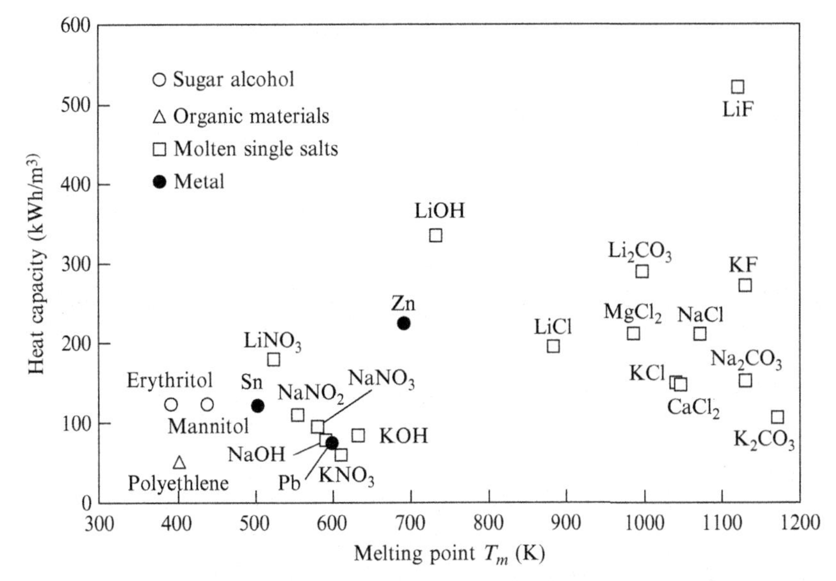

Fig. 3.5 Latent-heat thermal storage capacity of various PCMs. *(Courtesy of Hoshi A, Mills DR, Bittar A, Saitoh TS. Screening of high melting point phase change materials (PCM) in solar thermal concentrating technology based on CLFR. Sol Energy 2005;79:332–339).*

The product of density and the latent heat of a PCM is the energy storage capacity during the phase transition process. Hoshi et al. [31] collected some data for latent-heat energy storage capacity of PCMs for various materials (single or eutectic mixture) with melting points in the range from 300 K to 1200 K, as given in Figs. 3.5 and 3.6. It is seen that the melting point of a single material tends to increase in the order of nitrates, chlorides, carbonates, and fluorides.

Some detailed data of the surveyed PCMs are given in Tables 3.11–3.13. Table 3.11 shows that LiCl-LiOH (wt37%–wt67%) has the highest energy density in its phase change among the seven different types of PCMs. In Table 3.12, the melting temperatures of the thermal storage PCMs range from 220°C to 400°C, which matches the operating temperature of the steam Rankine cycle. It is obvious that $NaNO_3$, NaOH, KNO_3, NaCl-KCl (58:42%wt), and KOH are all good candidates for latent heat thermal energy storage (LHTES). Table 3.13 demonstrates some PCMs with potential for higher temperature applications, as needed for higher thermal efficiency in a CSP system. For example, NaF-MgF_2 (75:25 wt%) has a melting temperature at 650°C, which is good for high temperature LHTES applications.

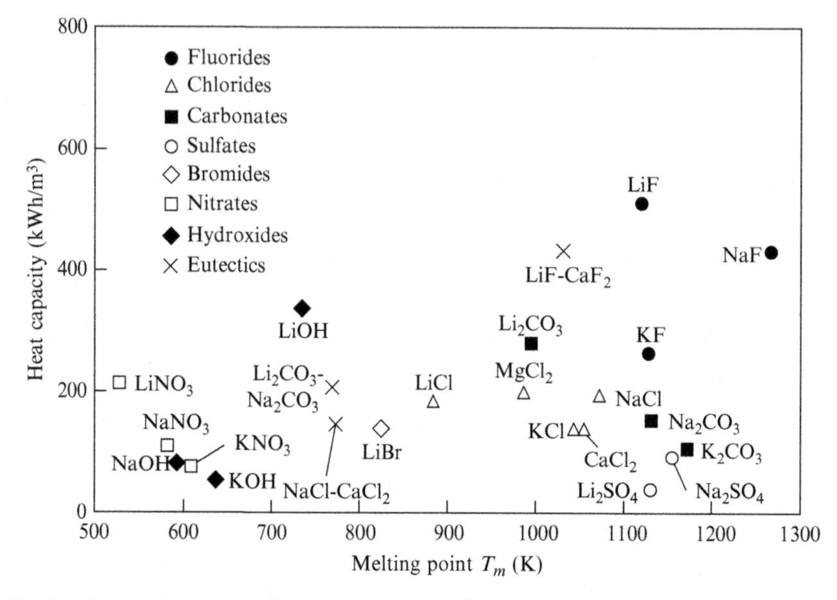

Fig. 3.6 Latent-heat thermal storage capacity of various molten salt PCMs. *(Courtesy of Hoshi A, Mills DR, Bittar A, Saitoh TS. Screening of high melting point phase change materials (PCM) in solar thermal concentrating technology based on CLFR. Sol Energy 2005;79:332–339).*

As shown in Fig. 3.7, there are two ways that PCM is packed in a thermal storage container. One is that the PCM is encapsulated into small spherical capsules and a packed bed of capsules is formed in a storage tank; the HTF flows through the packed bed. Another way is that the PCM is directly packed in a tank and pipes embedded in the PCM have fluid flow to deliver or withdraw the thermal energy from the PCM. The former approach has an effective heat transfer but often has challenges in the encapsulation techniques. In the latter approach, it is easy to pack the PCM, but low heat transfer performance is the disadvantage. If the PCM is encapsulated into a large number of capsules, either spheres or cylinders, the heat transfer area of the PCM dramatically increases, which offers much better heat transfer when HTF flows through the capsule packed bed. It has been reported that PCM stored in capsules with a radius of 10 mm offers a surface area of more than 300 m^2 per cubic meter. Depending on the techniques of encapsulation, capsules may have a size of several tens of millimeters to submillimeters. Metallic encapsulation is preferred if high temperature is the criterion; otherwise plastic encapsulation is commonly used for low temperature applications, as shown in Fig. 3.7.

Table 3.11 Inorganic PCMs with melting temperature between 100°C and 280°C

Compound (wt%)	Melting point (°C)	Latent heat (kJ/kg)	Density (kg/m^3)	Energy density (kJ/m^3)	Thermal conductivity (W/m K)
$MgCl_2 \cdot 6H_2O$ [32–35]	117	168.6	1450 (liquid @120°C) 1569 (solid, 20°C)	244,470 (liquid @120°C) 264,533 (solid @ 20°C)	0.570 (liquid @ 120°C) 0.694 (solid @ 90°C)
$NaNO_3$-KNO_3 (50:50%)[34,36]	220	100.7	1920	193,344	0.56
KCl-$ZnCl_2$ (68.1:31.9%) [34,35]	235	198	2480	491,040	0.8
LiCl-LiOH (37:67%) [32,33]	262	485	1550	751,750	1.10

Table 3.12 Inorganic PCMs with melting temperature between 280°C and 400°C

Compound (wt%)	Melting point (°C)	Latent heat (kJ/kg)	Density (kg/m^3)	Energy density (kJ/m^3)	Thermal conductivity (W/m K)
$ZnCl_2$ [34,36]	280	75	2907	218,025	0.5
$NaNO_3$ [34,36]	308	199	2257	449,143	0.5
NaOH [34,36]	318	165	2100	346,500	0.92
KNO_3 [34,36]	336	116	2110	244,760	0.5
NaCl-KCl (58:42%) [34,36]	360	119	2084.4	248,044	0.48
KOH [34,36]	380	149.7	2044	305,987	0.5

Table 3.13 Inorganic PCMs with melting temperature above 400°C

Compound (wt%)	Melting point (°C)	Latent heat (kJ/kg)	Density (kg/m^3)	Energy density (kJ/m^3)	Thermal conductivity (W/m K)
$MgCl_2$-NaCl (38.5:61.5%) [37–39]	435	351	2480	870,480	N/A
Na_2CO_3-Li_2CO_3 (56:44%) [39,40]	496	370	2320	858,400	2.09
NaF-MgF_2 (75:25%) [39,40]	650	860	2820	2,425,200	1.15
$MgCl_2$ [36]	714	452	2140	967,280	N/A
LiF-CaF_2 (80.5:19.5%) [39,41]	767	816	2390	1,950,240	1.70 (liquid) 3.80 (solid)
NaCl [39,42]	800	492	2160	1,062,720	5.0
Na_2CO_3 [39,41]	854	275.7	2533	698,348	2.0
K_2CO_3 [39,42]	897	235.8	2290	539,982	2.0

Because of the needs for packaging of the PCM and heat transfer enhancement, chemical compatibility or corrosion of containers or capsule shells are also important criteria for the candidacy of a PCM. Zalba et al. [36] summarized the advantages and disadvantages of organic and inorganic PCMs and concluded that inorganic PCMs have greater latent heat, but with worse corrosion resistance and weaker thermal stability. Organic PCMs have very low corrosion and much better thermal/chemical stability, but have a much lower latent heat and thermal conductivity.

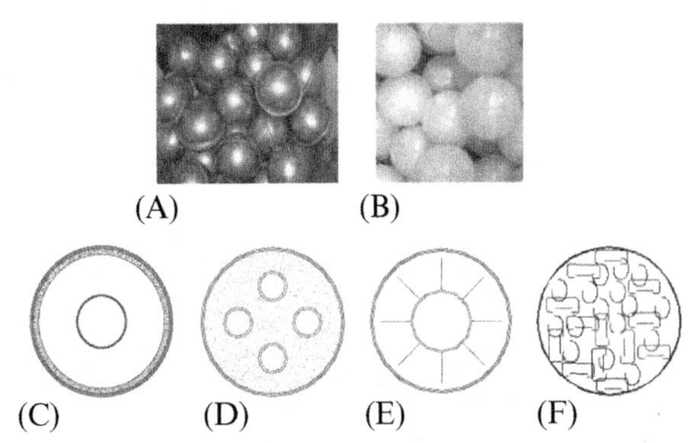

Fig. 3.7 PCM packing schemes. (A) High temperature PCM in metal shells; (B) low temperature PCM in plastic shells; (C) annulus PCM; (D) pipe-embedded PCM; (E) fins in PCM; (F) metal rings in PCM. *Reproduced from Agyenim F, Hewitt N, Eames P, Smyth M. A review of materials, heat transfer and phase change problem formulation for latent heat thermal energy storage systems (LHTESS). Renew Sust Energ Rev 2010; 14(2):615–628.*

In recent years, the technology of encapsulated PCMs for cold storage has developed significantly. Due to the encapsulation, the heat transfer surface area can be dramatically improved and thus the cold storage efficiency can be sufficiently high. Courtesy of Veerakumar and Sreekumar [43], the following tables summarize some low melting PCM materials (Tables 3.14–3.16).

Table 3.14 Low melting point organic PCM materials [43]

Materials	Melting point (°C)	Heat of fusion (kJ/kg)
Microencapsulated tetradecane	5.2	215
n-Tetradecane	5.5	215
Paraffin C14	5.5	228
Formic acid	7.8	247
Polyglycol E400	8	99.6
n-Pentadecane	10	193.9
Paraffin C15	10	205
Tetrabutyl ammonium bromide	10–12	193–199
Isopropyl palmitate	11	95–100
Isopropyl stearate	14–18	140–142
Propyl palmitate	16–19	186
Caprylic acid	16	150
Dimethyl sulphoxide	16.5	85.7
Paraffin C16	16.7	237.1

Continued

Table 3.14 Low melting point organic PCM materials [43]—cont'd

Materials	Melting point (°C)	Heat of fusion (kJ/kg)
Acetic acid	16.7	184
Polyethylene glycol 600	17–22	No data
Glycerine	17.9	198.7
n-Hexadecane	18	210–236
n-Heptadecane	19	240
Butyl stearate	19	140–200
Dimethyl sabacate	21	120–135
Octadecyl 3-mencaptopropylate	21	143
Paraffin C17	21.7	213
Paraffin C16–C18	20–22	152
Paraffin C13–C24	22–24	189
Ethyl palmitate	23	122
Lactic acid	26	184
1-Dodeconol	26	200
Octadecyl thioglyate	26	90
Vinyl sterate	27–29	122
Paraffin C18	28	244
n-Octadecane	28–28.1	245
Methyl sterate	29	169

Table 3.15 Low melting point inorganic PCM materials [43]

Materials	Melting point (°C)	Heat of fusion (kJ/kg)
H_2O	0	333
H_2O + polyacrylamide	0	295
$K_2HPO_4 \cdot 6H_2O$	4	109
$LiClO_3 \cdot 3H_2O$	8	155–253
$ZnCl_2 \cdot 3H_20$	10	No data
$K_2HPO_4 \cdot 6H_2O$	13	No data
$NaOH \cdot (3/2\ H_2O)$	15–15.4	No data
NaOH	16	200
$Na_2CrO_4 \cdot 10H_2O$	18	No data
$KF \cdot 4H_2O$	18.5	231
$Na_2SO_4 \cdot 10H_2O$	21	198
$FeBr_3 \cdot 6H_2O$	21	105
$Mn(NO_3)_2 \cdot 6H_2O$	25.8	125.8
$CaCl_2 \cdot 6H_2O$	29	190.8
$CaCl_2 \cdot 12H_2O$	29.8	174
$LiNO_3 \cdot 2H_2O$	30	296
$LiNO_3 \cdot 3H_2O$	30	189/296

Table 3.16 Low melting point commercial PCM materials [43]

Name	Material	Melting point (°C)	Heat of fusion (kJ/kg)
RT3	Paraffin	3	198
RT4	Paraffin	4	182
RT5	Paraffin	5	198
RT6	Paraffin	6	175
MPCM(6)	Paraffin	6	157–167
ClimSel C7	Salt solution	7	130
E17	Salt hydrate	17	143
E19	Salt hydrate	19	146
RT20	Paraffin	20	140
Emerest 2325	Fatty acid	20	134
Emerest 2326	Fatty acid	20	139
FMC	Paraffin	20–23	130
RT21	Paraffin	21	134
ClimSel C21	Salt solution	21	122
S21	Salt hydrate	21	170
E21	Salt hydrate	21	150
SP22 A17	Blend	22	180
A22	Paraffin	22	145
ClimSel C23	Salt hydrate	23	148
S23	Salt hydrate	23	175
GM	Ceramic +paraffin	23.5–24.9	41.9
ClimSel C24	Salt hydrate	24	216
GR25	Granule	23–25	No data
A24	Paraffin	24	145
SP22A4	Blend	24	165
ClimSel C24	Salt solution	24	180
S25	Salt hydrate	25	180
SP25A8	Blend	25	180
LatestTM25T	Salt hydrate	24–26	175
RT26	Paraffin	24–28	131
A26	Paraffin	26	150
RT27	Paraffin	27	184
STL27	Salt hydrate	27	213
S27	Salt hydrate	27	183
LatestTM29T	Salt hydrate	28–29	175
E30	Salt hydrate	30	201
ClimSel C32	Salt hydrate	32	212

3.6 DEGRADATION OF THERMAL STORAGE MATERIALS AND CORROSION OF METALS

Thermal storage material (solid or liquid), heat transfer fluids, and materials of the system accommodating the thermal storage materials may experience degradation due to unwanted chemical reactions at exceedingly high temperatures, fatigue from thermal cycles, problems due to nonuniform thermal expansions, and corrosion of materials.

When mineral or synthetic oils have too high a temperature, they chemically decompose and may release gases that cause pressure rise in a closed system. They have the advantage that oils do not corrode metallic containers and pipes.

Molten salt fluids as thermal storage materials may have chemical reactions when their temperature exceeds a safety limit accidentally, due to localized overheating. For example, HITEC heat transfer salt is thermally stable up to 454°C; the maximum recommended operating temperature is only 538°C, although the salt can be used for a short time at temperatures above 550°C. When HITEC is used at temperatures beyond 454°C in a closed system, it undergoes a slow thermal breakdown (endothermic) of the nitrite to nitrate, alkali metal oxide, and nitrogen [44]:

$$5NaNO_2 \rightarrow 3NaNO_3 + Na_2O + N_2 \uparrow \tag{3.34}$$

The loss of nitrite in the eutectic salt will cause an increase of the freeze point, which is unwanted in CSP systems.

When HITEC is used in an open system, in contact with the air, and in the higher operating range between 454 and 538°C, the nitrite is slowly oxidized by atmospheric oxygen:

$$NaNO_2 + \tfrac{1}{2}O_2 \leftrightarrow NaNO_3 \tag{3.35}$$

which is a process also applicable to other nitrate salts, such as KNO_2 or $LiNO_2$. The loss of the nitrite component will cause an increase of the freeze point of the salt.

Other minor reactions that gradually alter the compositions of HITEC and Solar Salt include: (1) the absorption of carbon dioxide to form carbonates which may precipitate, and (2) the absorption of water vapor to form alkali metal hydroxides. Therefore, to eliminate all of these reactions that cause the raising of the freezing point, nitrogen blanketing of the molten salts is typically applied.

The corrosion rates of HITEC salts on several types of metals used for containers and pipes are listed in Table 3.17, which is based on the data

Table 3.17 Corrosion rate of HITEC on several types of metals

	Corrosion rate μm/year						
					570°C		
Metals	**322°C**	**418°C**	**454°C**	**538°C**	**First period**	**Second period**	**593°C**
Steel-open hearth (ASTM A273, A274)	–	–	91	305–610	–	–	3048–15,240
Alloy steel 15%–16% chromium iron stainless steel	–	–	–	0	–	–	–
Type 304	–	–	–	213	–		–
Type 304 L	–	–	–	1829	–		–
Type 309 (annealed)	6	3	0	–	335	195	–
Type 309 Cb	–	–	–	–	475	287	–
Type 310	–	–	–	–	357	235	–
Type 316	–	–	–	0	–	–	–
Type 321	–	–	–	–	338	171	–
Type 347	–	–	–	122	332	207	–
Type 446	–	–	–	–	445	219	–
Inconel	–	–	–	0	466	460	–
Carpenter 20	–	–	–	–	296	180	–
Hastelloy B	34	1	–	–	–	–	–
Monel	–	–	–	30	–	–	–
Bronze	18	24	30	–	–	–	–
Phosphonized Admiralty	18	15	30	–	–	–	–
							–
Copper	–	–	–	9144	–	–	–
Nickel	–	–	–	–	–	–	762

reported in Ref. [44]. The corrosion rate increases dramatically when the temperature reaches 538°C.

For molten chloride salts, water and oxygen contamination is a big issue. When water reacts with $ZnCl_2$ or $MgCl_2$, it will create the gas HCl, which is corrosive to metals. Since chloride salts are projected to be used for higher temperature at the level of 800°C, only nickel-based high temperature alloys such as Hastelloys and Haynes are considered to be suitable metals to resist corrosion.

The typical problem in solid thermal storage materials such as concrete is the cracking, as shown in Fig. 3.8. The cracks are created due to moisture-created pressure when the concrete is heated, or due to the mismatch of thermal expansion of the concrete and metal pipes in the thermal cycles of heat charge and discharge, which will cause significant degradation of thermal conduction in the material and thus results in reduction of the thermal efficiency of the system. Filling thermal conductive liquid in the cracks or making the thermal expansion matched between the metal tubes and concrete materials can be the solutions.

Thermal ratcheting [45] is the issue in thermal storage systems that use packed beds of pebbles or particles (such as rocks, sands, etc.). The thermal ratcheting is a process caused by the phenomenon in which the tank wall of the packed bed for thermal storage expands more than the packed material when heated. This mismatched expansion may cause a gap between the tank wall and the packed bed material. This may lead to the falling of some of the

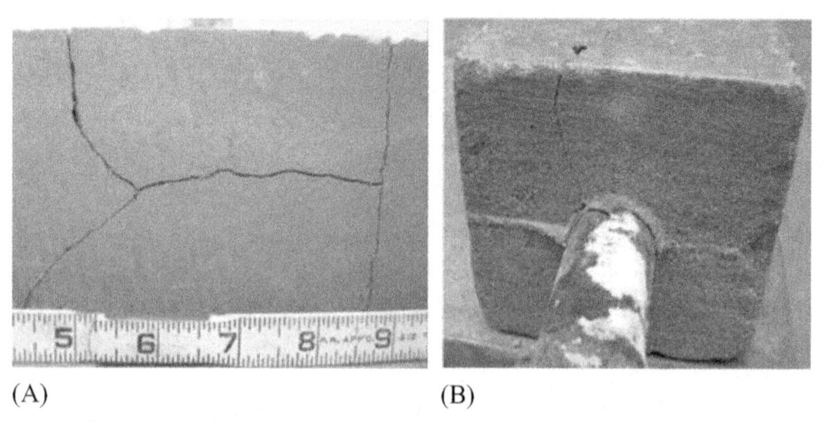

Fig. 3.8 Concrete cracking due to thermal cycles. (A) cracks inside concrete; (B) cracks between fluid tubes and concrete. *(Courtesy of reference Skinner JE, Strasser MN, Brown BM, Selvam RP. Testing of high-performance concrete as a thermal energy storage medium at high temperatures. J Sol Energy Eng 2014; 136(2):021004).*

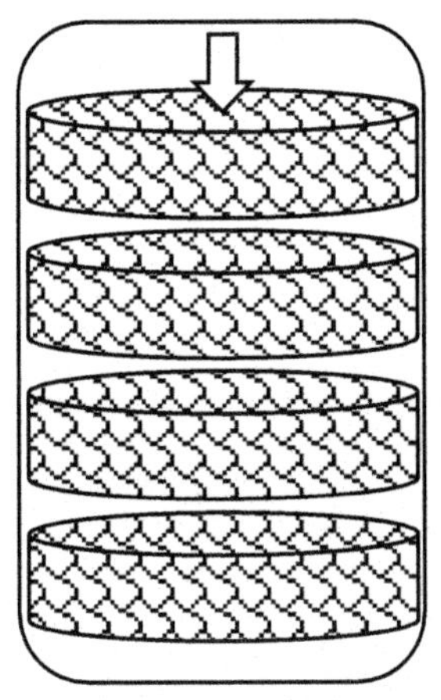

Fig. 3.9 Solid material to be packed into separately framed zones in one tank.

rocks/particles into the gap between the tank wall and the originally packed rocks/particles, which makes the tank wall not able to contract back to its original diameter when cooled down. The progress of this process in the thermal cyclic operation will cause the tank diameter to become larger and larger and the building up of higher and higher stress, which will lead to catastrophic failure of the wall material.

One remedy for the thermal ratcheting effect is to have a good packaging scheme for the solid particles or rocks in the packed bed. For example, in Fig. 3.9 the entire packed bed is divided into multiple zones and each zone is packed in a short height with rocks or particles frame-restricted, and thus there is less chance for the rocks to fall into the gap between the tank wall and the packed bed.

REFERENCES

[1] Fletcher AE. Solar thermal processing: a review. J Sol Energy Eng 2001;123:63–74.

[2] Stekli J, Irwin L, Pitchumani R. Technical challenges and opportunities for concentrating solar power with thermal energy storage. ASME J Therm Sci Eng Appl 2013;5:021011-1.

[3] XCELTHERM®. 600 Engineering properties. Radco Industries. http://www.radcoind.com/; 2016 [accessed 11.11.16].

[4] Canada S, Brosseau DA, Price H. Design and construction of the APS 1 MWe parabolic trough power plant, In: ASME conference proceedings; 2006. p. 91–8.

[5] Valmiki MM, Li P, Heyer J, Morgan M, Albinali A, Alhamidi K, et al. A novel application of a Fresnel lens for a solar stove and solar heating. Renew Energy 2011;36(5): 1614–20.

[6] Lang C, Lee B. Heat transfer fluid life time analysis of diphenyl oxide/biphenyl grades for concentrated solar power plants. International Conference on Concentrating Solar

Power and Chemical Energy Systems, SolarPACES 2014. Energy Procedia 2015;69: 672–80.
[7] Therminol VP-1. http://www.solutia.com. [accessed 28.2015].
[8] Sohal MS, Ebner MA, Sabharwall P, Sharpe P. Engineering database of liquid salt thermophysical and thermochemical properties. Report # INL/EXT-10-18297, Idaho Falls, ID: Idaho National Laboratory; 2010.
[9] Bradshaw RW, Siegel NP. Molten nitrate salt development for thermal energy storage in parabolic trough solar power systems. In: Proceedings of ES2008, energy sustainability 2008, August, 10–14, Jacksonville, FL; 2008. p. 1–7.
[10] Baraka A, Abdel-Rohman AI, El Hosary AA. Corrosion of mild steel in molten sodium nitrate-potassium nitrate eutectic. Br Corros J 1976;11(1):43–6.
[11] Serrano-López R, Fradera J, Cuesta-López S. Molten salts database for energy applications. Chem Eng Process 2013;73:87–102.
[12] Li P, Molina E, Wang K, Xu X, Dehghani G, Kohli A, et al. Thermal and transport properties of NaCl–KCl–$ZnCl_2$ eutectic salts for new generation high-temperature heat-transfer fluids. J Sol Energy Eng 2016;138:054501.
[13] Thermophysical and Electric Properties. Chapter 2, Handbook on Lead-bismuth Eutectic Alloy and Lead Properties, Materials Compatibility, Thermal-hydraulics and Technologies. Boulogne-Billancourt: OECD Nuclear Energy Agency (NEA).
[14] Low Prandtl number thermal-hydraulics. Chapter 10, Handbook on Lead-bismuth Eutectic Alloy and Lead Properties, Materials Compatibility, Thermal-hydraulics and Technologies. Boulogne-Billancourt: OECD Nuclear Energy Agency (NEA). p. 48–57.
[15] Boerema N, Morrison G, Taylor R, Rosengarten G. Liquid sodium versus Hitec as a heat transfer fluid in solar thermal central receiver systems. Sol Energy 2012;86 (9):2293–305.
[16] Pacio J, Singer C, Wetzel T, Uhlig R. Thermodynamic evaluation of liquid metals as heat transfer fluids in concentrated solar power plants. Appl Therm Eng 2013; 60(1–2):295–302.
[17] Pacio J, Wetzel T. Assessment of liquid metal technology status and research paths for their use as efficient heat transfer fluids in solar central receiver systems. Sol Energy 2013;93:11–22.
[18] Sang LX, Cai M, Ren N, Wu YT. Study on modification of ternary carbonates with additives. Energy Procedia 2015;69:1023–8.
[19] Wu Y-t, Ren N, Wang T, Ma C-f. Experimental study on optimized composition of mixed carbonate salt for sensible heat storage in solar thermal power plant. Sol Energy 2011;85:1957–66.
[20] Bejan A. Entropy generation minimization—The method of thermodynamic optimization of finite-size systems and finite-time processes. Boca Raton, FL: CRC Press; 1995.
[21] Bejan A. Advanced engineering thermodynamics. 3rd ed. Hoboken, NJ: Wiley; 2006.
[22] Incropera FP, DeWitt DP. Introduction to heat transfer. 4th ed. New York, NY: Wiley; 2002.
[23] Gnielinski V. Int Chem Eng 1976;16:359.
[24] Kays WM, Crawford ME. Convective heat and mass transfer. New York, NY: McGraw-Hill; 1980.
[25] Petukhov BS. In: Irvine TF, Hartnett JP, editors. Advances in heat transfer, 6: New York, NY: Academic Press; 1970.
[26] Notter RH, Sleicher CH. A Solution to the turbulent graetz-problem III fully developed and entrance region heat transfer rates. Chem Eng Sci 1972;27:2073–93.
[27] Li P-W, Van Lew J, Karaki W, Lik Chan C, Stephens J, O'Brien JE. Transient heat transfer and energy transport in packed bed thermal storage systems. In: dos Santos Bernardes MA, editor. Developments in heat transfer. Rijeka: InTech; 2011. [chapter 20], 978-953-307-569-3.

[28] Hamilton RL, Crosser OK. Thermal conductivity of heterogeneous two component systems. Ind Eng Chem Fundam 1962;1(3):187–91.
[29] Gonzo EE. Estimating correlations for the effective thermal conductivity of granular materials. J Chem Eng 2002;90:299–302.
[30] Han J, Xu B, Li P, Kumar A, Yang Y. Experimental study of a novel thermal storage system using sands with high-conductive fluids occupying the pores. In: Proceedings of the ASME 2014 international mechanical engineering congress & exposition, IMECE 2014-38999, Nov. 14–20, Montreal, Quebec, Canada; 2014.
[31] Hoshi A, Mills DR, Bittar A, Saitoh TS. Screening of high melting point phase change materials (PCM) in solar thermal concentrating technology based on CLFR. Sol Energy 2005;79:332–9.
[32] Dincer I, Rosen MA. Thermal energy storage, systems and applications. Chichester: John Wiley & Sons; 2002.
[33] Lane GA. Low temperature heat storage with phase change materials. Int J Ambient Energy 1980;1:155–68.
[34] Pincemin S, Olives R, Py X, Christ M. Highly conductive composites made of phase change materials and graphite for thermal storage. Sol Energy Mater Sol Cells 2008;92:603–13.
[35] Farkas D, Birchenall CE. New eutectic alloys and their heats of transformation. Metall Trans A 1985;16A:324–8.
[36] Zalba B, Marı'n JM, Cabeza LF, Mehling H. Review on thermal energy storage with phase change: materials, heat transfer analysis and applications. Appl Therm Eng 2003;23(3):251–83.
[37] Marianowski LG, Maru HC. Latent heat thermal energy storage systems above 450°C, In: Proceedings of 12th intersociety energy conversion engineering conference; 1977. p. 555–66.
[38] Maru HC, Dullea JF, Kardas A, Paul L, Marianowski LG, Ong E, et al. Molten salts energy storage systems. Final Report, Chicago, IL: Institute of Gas Technology; 1978.
[39] Agyenim F, Hewitt N, Eames P, Smyth M. A review of materials, heat transfer and phase change problem formulation for latent heat thermal energy storage systems (LHTESS). Renew Sust Energ Rev 2010;14(2):615–28.
[40] Birchenall CE, Riechman AF. Heat storage in eutectic alloys. Metall Trans A 1980;11A (8):1415–20.
[41] Gong Z, Mujumdar AS. Finite-element analysis of cyclic heat transfer in a shell and tube latent heat energy storage exchanger. Appl Therm Eng 1997;17(4):583–91.
[42] Pilkington Solar International. Survey of thermal storage for parabolic trough power plants; 2002, Report NREL1EC24.
[43] Veerakumar C, Sreekumar A. Phase change material based cold thermal energy storage: materials, techniques and applications—a review. Int J Refrig 2016;67:271–89.
[44] Hitec®. Heat transfer salt. Document issues by Coastal Chemical Co., L.L.C.
[45] Motte F, Bugler-Lamb SL, Falcoz Q, Py X. Numerical study of a structured thermocline storage tank using vitrified waste as filler material. Energy Procedia 2014;49: 935–44.

FURTHER READING

[1] Salunkhe PB, Shembekar PS. A review on effect of phase change material encapsulation on the thermal performance of a system. Renew Sust Energ Rev 2012;16(8):5603–16.
[2] Skinner JE, Strasser MN, Brown BM, Selvam RP. Testing of high-performance concrete as a thermal energy storage medium at high temperatures. J Sol Energy Eng 2014;136(2): 021004.

CHAPTER 4

Mathematical Models and Numerical Solutions for Thermal Storage Processes

Contents

Abstract

This chapter presents information on mathematical models for thermal storage, covering the establishing of proper governing equations to mathematically follow the energy conservation principles for "control volumes" in a thermal storage tank when heat is charged or withdrawn; deciding the boundary condition requirements for the governing equations; and discovering the most efficient mathematical method to solve the governing equations with accuracy, so that the temperatures of fluid and solid media can be determined at any location in the tank at any time. The models address several

Thermal Energy Storage Analyses and Designs
http://dx.doi.org/10.1016/B978-0-12-805344-7.00004-3

configurations of thermal storage systems, including single fluid thermal storage, dual media (fluid and packed bed) sensible thermal storage, and dual-media PCM-based latent heat thermal storage. In the dual-media sensible thermal storage, both configurations of solid particle packed bed and integrated solid with fluid pipes passing through are considered and discussed. The transient heat transfer in a packed bed is typically assumed uniform in the radial direction in a storage tank and thus one- or two-dimensional governing equations are sufficient to describe the problems. For convenient design analysis, the models described in the following paragraphs start from a one-dimensional model which has been proven to have no sacrifice of accuracy. Models in three dimensions considering nonuniform flow are briefly introduced at the end of the chapter.

Keywords: Ideal thermal storage, Energy delivery efficiency, Thermal energy storage modeling, PCM, Enthalpy method, Method of characteristics

Nomenclature

a_f the cross-sectional area of a storage tank (m^2)
Bi Biot number ($= L_p h/k_s$)
C heat capacity (J/kg °C)
d_r nominal diameter of a single filler "particle" (rocks) (m)
$\bar{h}$ enthalpy (J/kg)
H heat transfer coefficient (W/m^2 °C)
h_{eff} effective heat transfer coefficient (W/m^2 °C)
H length or height of a storage tank (m)
H_{CR} a dimensionless parameter, Eq. (4.12)
k thermal conductivity (W/m °C)
L latent heat of fusion of storage PCM
L_p characteristic length of particles for Biot number
$\dot{m}$ mass flow rate (kg/s)
N number of tubes for HTF in a storage tank
Pr Prandtl number
R radius of the storage tank (m)
Re modified Reynolds number for porous media
S_s surface area of filler material per unit length of the storage tank (m)
t time (s)
T temperature (°C)
T_H high temperature of fluid from solar field (°C)
T_L low temperature of fluid from power plant (°C)
U fluid velocity in the axial direction in the storage tank (m/s), Eq. 4.4
V volume (m^3)
V_f volume of fluid per unit length of the storage tank (m^2)
V_s volume of filler material per unit length of the storage tank (m^2)
z location of a fluid element along the axis of the tank (m)

Greek Symbols

ε porosity of packed bed in a storage tank
η_s thermal storage efficiency

η_s dimensionless enthalpy of the encapsulated PCM
ν kinematic viscosity (m^2/s)
Π dimensionless charge or discharge time
τ_r a dimensionless parameter, see 4.9
P density (kg/m^3)
Θ dimensionless temperature
ϑ nodes of the numerical grid

Subscript

c energy charge process
d energy discharge process
f thermal fluid
final equilibrium temperature after each charge or discharge
ref a required reference value
s filler material (rocks), the primary thermal storage material
z location along the axis of the tank

Superscript

* dimensionless values

The most important goal of investigation and studies of thermal energy storage is to design the size of tanks for thermal storage of a certain energy demand over a desired period of time, and also to predict and understand the variation of fluid temperatures discharged from a storage tank. To accomplish the goal, we must rely on a fundamental study of the energy flow and the heat transfer between the heat transfer fluid and solid materials in a packed bed, if dual-media thermal storage is adopted.

The content of this section is a presentation of the mathematical modeling of such systems, including:

(1) Establishing proper governing equations to mathematically follow the energy conservation principles for "control volumes" in a thermal storage tank when heat is charged or withdrawn.
(2) Deciding the requirement of boundary conditions for the governing equations.
(3) Finding out the most efficient mathematical method to solve the governing equations with accuracy so that the temperatures of the fluid and solid media can be determined at any location in the tank at any time.

The models address several configurations of the thermal storage systems presented in Chapter 2, including single fluid thermal storage, dual-media (fluid and packed bed) sensible thermal storage, and dual-media phase change material (PCM) based latent heat thermal storage. In the dual-media sensible thermal storage, both configurations of solid particle packed bed and integrated solid with fluid pipes passing through are considered and discussed.

The transient heat transfer in a packed bed is typically assumed uniform in the radial direction in a storage tank and thus one- or two-dimensional governing equations are sufficient to describe the problems. For convenient design analysis, the models described in the following subsections start from a one-dimensional model which has been proven to have no sacrifice of accuracy. Models in three dimensions considering nonuniform flow are briefly introduced at the end of the chapter.

4.1 IDEAL THERMAL STORAGE USING LIQUID ALONE

In an ideal thermal storage system, high temperature heat transfer fluid (HTF) is stored, and when it is withdrawn there should be no temperature degradation. Such a system requires that there be no heat loss and no heat transfer when the HTF is stored in or withdrawn from a tank.

Contingent upon the thermal insulation being perfectly maintained, the two-tank HTF storage system in Fig. 2.1 can operate like an ideal thermal energy storage system. It has been discussed before that a two-tank storage system can be replaced by a single-tank storage system, as shown in Fig. 2.2, in which a stratification of fluid (hot on top of cold), or a thermocline mechanism, must be maintained. However, even if the thermocline is maintained, the heat conduction between hot fluid and cold fluid may cause a temperature drop in the hot fluid, which will not allow for ideal thermal storage performance. A modification proposed by the current authors [1] uses a thermal insulation baffle in the single tank, which separates the hot fluid from the cold fluid, as shown in Fig. 4.1. In this system, if the floating thermal insulation baffle prevents heat conduction from the hot fluid to the cold fluid, an ideal thermal storage performance can be achieved. With regard to both cost reduction and energy storage performance, the single tank with floating thermal insulation baffle is the ideal thermal storage system considered in this chapter.

Whereas physically an ideal thermal storage system has the clearly identifiable features previously detailed, mathematically it should be described as a system that has an energy storage efficiency of 1.0. With this in mind, the following definition of thermal energy delivery efficiency is adopted for thermal energy storage systems:

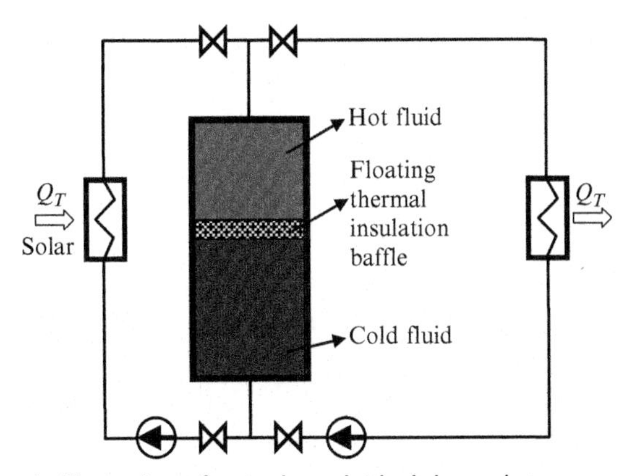

Fig. 4.1 Schematic illustration of a single tank ideal thermal storage system.

$$\eta = \frac{\int_0^{t_{ref,\ discharge}} \left[T_{f\,(z=H,t)} - T_L \right] dt}{(T_H - T_L) \cdot t_{ref,\ discharge}} \tag{4.1}$$

where z is the vertical coordinate of the tank and H is the height of the tank. In adopting this definition, we assume that the average heat capacity, Cp, and the mass flow rates of the HTF for the charging and discharging processes are the same. The integration on the numerator of Eq. (4.1) is the energy discharge in an actual process, and the value in the denominator is the ideal energy discharge.

For the ideal thermal storage system, it is assumed that the temperature of the hot fluid in a charging process is kept at constant T_H; and in the discharging process the discharged fluid maintains a constant temperature of T_H as well. After releasing heat in a heat exchanger, the fluid returns to the bottom of the storage tank at a constant temperature, T_L. To substitute these conditions from the ideal thermal storage system into Eq. (4.1), the fluid temperature $T_{f\,(z=H,t)}$ during the discharge process in a time period of 0 to $t_{ref,\ discharge}$ should be equal to the high temperature, T_H. This will make the energy delivery efficiency equal to $\eta = 1.0$ for the ideal thermal storage system.

In a real thermal energy storage system, such as the systems shown in Fig. 2.3, it is easy to understand that when cold fluid is pumped into the tank from the bottom, it will extract heat from the solid thermal storage material and be warmed up when it flows out of the tank. However, after a certain time, the cold fluid going into the tank may not be heated up sufficiently before it flows out from the top of the tank. Unfortunately, this temperature degradation is inevitable due to the heat transfer between the solid thermal storage material and the HTF, even if initially the solid thermal storage material is fully charged, or its temperature is exactly equal to T_H.

Considering the need for an HTF in a power plant, it is always important that, during the required operational period of time $t_{ref,discharge}$, the temperature of the HTF have minimal or no degradation from the temperature at which the fluid is stored. To meet this requirement in an actual thermocline storage system, one needs to first store a sufficient amount of energy (more than the ideal amount) in the tank. This requires a storage tank having a sufficiently large thermal energy storage capacity as well as a sufficiently long charge time that allows heat to be charged to the tank. Giving this requirement as a mathematical expression, it is:

$$\left\{\left[\rho_s C_s(1-\varepsilon)+\rho_f C_f \varepsilon\right] V_{real}\right\} > \left[\left(\rho_f C_f\right) V_{ideal}\right] \tag{4.2}$$

In engineering reality, one needs to know specifically how large the real thermal storage volume, V_{real}, is and how long a charging time is needed, if the assumed operation time period of a power plant is $t_{ref,discharge}$. This must be addressed through mathematical analysis.

In the following section, the modeling of the heat transfer and energy transport between the solar thermal storage material and the HTF is described. The goal of the modeling analysis is to predict the size of the storage tank and the period of time required to charge the tank for a given subsequent period of heat discharge from the system, within which minimal or no temperature degradation must be maintained.

4.2 ONE-DIMENSIONAL MODEL FOR DUAL-MEDIA PACKED-BED SENSIBLE THERMAL STORAGE

In Chapter 2, the various types of thermal storage systems were discussed. The idea of a thermocline thermal storage system, with a medium and an HTF, is to deposit the thermal energy from the hotter temperature HTF. The stored thermal energy is retrieved by passing a cooler HTF through the storage system. The heat transfer involved is the convection within the HTF and conduction in the storage medium. This is in fact a conjugate heat transfer problem. The convection and conduction are to be solved simultaneously. They are coupled through the interface via temperature and heat flux matching. The fluid flow, on the other hand, has to satisfy no slip and no penetration conditions along the interface. Furthermore, boundary conditions on the outer physical boundary have to be enforced, depending on the situation.

The numerical solution of the mathematical model for the conjugate heat transfer can be quite time consuming. To reduce the computational time, simplifications are made. Furthermore, the simplified models allow thorough parametric studies. Since the computational times are short, this can be used as a fast and accurate design tool. In this section, a one-dimensional model is presented for a thermocline thermal storage system with packed bed for sensible energy storage. The heat transfer between the packed bed and HTF is modeled by Newton's cooling law.

Fig. 4.2 shows the schematic diagram of the one-dimensional sensible energy storage model. A differential control volume of thickness dz is selected in the packed bed. For convenience in analysis, the positive direction of coordinate z is set to be in the direction of the fluid flow. In the energy-charging process, hot fluid flows into the tank from the top, and thus $z=0$ is at the top of the tank. During the heat-discharging process, cold fluid flows into the tank from the bottom to extract heat from the solid material, and this makes $z=0$ at the bottom of the tank.

The following assumptions are made to reasonably simplify the analysis of the heat transfer between the HTF and the solid packing material:

(1) There is a uniform radial distribution of the fluid flow and filler material throughout the storage tank. This allows the model to be one dimensional, only in the z direction.

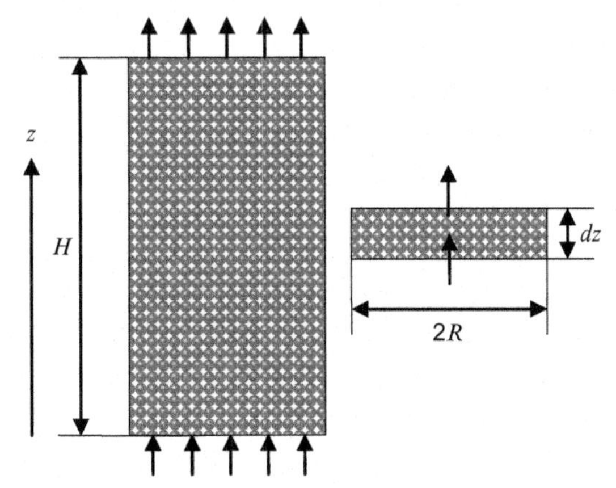

Fig. 4.2 Schematic of a packed-bed thermal storage system and a control volume for analysis.

(2) The particles of filler material have only point contact and therefore heat conduction between filler material is negligible.
(3) The heat conduction in the axial direction in the fluid is negligible compared to the convective heat transfer.
(4) The lumped heat capacitance method is applied to the transient heat conduction in the filler material (particles of size of 0.1–5.0 cm in nominal diameter). When this method is inadequate, due to the large size of the solid filler material, a *modified lumped capacitance method* will be used, which introduces a modified heat transfer coefficient for the convection heat transfer between the fluid and the solid filler material.
(5) There is no heat loss from the storage tank to the surroundings. This assumption applies to both the processes of energy charge and discharge, as well as the resting time between a charge and a discharge.

The assumption (3) is valid when the Peclet number (=RePr) in the HTF is sufficiently large, which is satisfied for most thermal energy storage applications [2]. Assumption (4) is valid when the Biot number $\left(= hL_p/k_s\right)$ for the thermal storage material is sufficiently small [3]. If the Biot number is large, a correction to the heat transfer accounting for the effects of an internal temperature gradient in the filler material will be considered. Heat loss from a thermal storage tank is inevitable and should also be considered [4]. However, from the design point of view, one needs to first decide the dimensions of the storage tank in order to find the heat loss. To compensate for the heat loss from the tank, a larger volume heat storage tank and a longer heat charge period may be adopted. A simple way of refining this design is to increase both the heat charge time and tank size with a factor that is equal to the ratio of heat loss versus the projected heat delivery. To focus on the main issues, the current work determines the dimensions of a storage tank without considering heat loss. The assumption of no heat loss to the surroundings also provides a basis for using the results from a heat charge process as the initial condition of the following discharge process, and vice versa. By using the end results of one process as the initial conditions of the following process, multiple cyclic energy charges and discharges in the actual operation can be simulated relatively easily.

Based upon these modeling assumptions (1), the cross-sectional area of the tank seen by the fluid flow is assumed constant at all locations along the axis of the tank, which gives:

$$a_f = \varepsilon \pi R^2 \tag{4.3}$$

The thermal energy balance of the fluid in the control volume dz is:

$$\rho_f \varepsilon \pi R^2 U(\hbar_z - \hbar_{z+dz}) + hS_s(T_s - T_f)dz = \rho_f C_f \varepsilon \pi R^2 dz \frac{\partial T_f}{\partial t} \tag{4.4}$$

where the parameter S_s denotes the heat transfer surface area between the filler material and the HTF per unit length of the tank; U is the actual fluid velocity in the packed bed:

$$U = \frac{\dot{m}}{\rho_f a_f} \tag{4.5}$$

The heat transfer coefficient h in Eq. (4.4) is for the convection between the HTF and the packing material. It can be different depending on the flow, packing condition of thermal storage material, fluid properties, and the interaction between packed material and HTF, such as in the schematic shown in Fig. 2.5. Detailed discussions of S_s and h are presented following the modeling work.

Using the definition of enthalpy change and a Taylor's series expansion, $\hbar_{z+dz} - \hbar_z = C_f(\partial T_f/\partial z)dz$, the energy balance equation for the HTF becomes:

$$\frac{hS_s}{\rho_f C_f \varepsilon \pi R^2}(T_s - T_f) = \frac{\partial T_f}{\partial t} + U\frac{\partial T_f}{\partial z} \tag{4.6}$$

We introduce the following dimensionless variables:

$$\theta_f = (T_f - T_L)/(T_H - T_L) \tag{4.7a}$$

$$\theta_s = (T_s - T_L)/(T_H - T_L) \tag{4.7b}$$

$$z^* = z/H \tag{4.7c}$$

$$t^* = t/(H/U) \tag{4.7d}$$

The dimensionless governing equation for the HTF is finally reduced to:

$$\frac{\partial \theta_f}{\partial t^*} + \frac{\partial \theta_f}{\partial z^*} = \frac{1}{\tau_r}(\theta_s - \theta_f) \tag{4.8}$$

where

$$\tau_r = \frac{U\rho_f C_f \varepsilon \pi R^2}{H \quad hS_s} = \frac{C_f \dot{m}}{HhS_s} \tag{4.9}$$

The boundary condition for Eq. (4.8) is from the fluid inlet temperature, while the initial condition is the temperature distribution in a tank before a charge or a discharge starts.

For the energy balance of the filler material in a control volume dz as shown in Fig. 4.2, it is understood that the filler material delivers or takes heat to or from the passing fluid at the cost of a change in the internal energy of the filler. The energy balance equation is:

$$hS_s\left(T_s - T_f\right)dz = -\rho_s C_s(1-\varepsilon)\pi R^2 dz \frac{\partial T_s}{\partial t} \tag{4.10}$$

By substituting in the dimensionless variables given in Eq. (4.7a)-(4.7d), the preceding governing equation for filler material is reduced to:

$$\frac{\partial \theta_s}{\partial t^*} = -\frac{H_{CR}}{\tau_r}\left(\theta_s - \theta_f\right) \tag{4.11}$$

where

$$H_{CR} = \frac{\rho_f C_f \varepsilon}{\rho_s C_s(1-\varepsilon)} \tag{4.12}$$

The boundary condition for Eq. (4.11) is the fluid inlet temperature (varies with time) and also the solid temperature at the inlet point (directly obtained from Eq. 4.11 for the inlet point alone).

In the energy charge and discharge processes, the filler material and HTF will have a temperature difference at any local location. Once the fluid comes to rest upon the completion of a charge or discharge process, the fluid will equilibrate with the local filler material to reach the same temperature, T_{final}. The energy balance of this situation at a local location is:

$$\varepsilon\rho_f C_f T_{f-initial} + (1-\varepsilon)\rho_s C_s T_{s-initial} = \varepsilon\rho_f C_f T_{final} + (1-\varepsilon)\rho_s C_s T_{final} \tag{4.13}$$

Here, the initial temperatures of primary thermal storage material and HTF are from the results of their respective charge or discharge processes. The final temperatures of the storage material and the fluid are the same after their thermal equilibrium is reached.

According to the assumption of no heat loss from the storage tank, it can be seen that the equilibrium temperature at the end of one process (charge or discharge) will necessarily be the initial condition of the next process in the cycle. This connects the discharge and charge processes so that overall periodic results can be obtained.

The initial temperatures of filler material and fluid in the storage tank should be known. Also, the inlet fluid temperature is known as a basic boundary condition, with which the filler temperature at inlet location $z=0$ can be easily solved mathematically from Eq. (4.11).

A joint solution of Eqs. (4.8) and (4.10) simultaneously is needed in order to find the temperature of both fluid and solid at a time and location.

4.2.1 Energy Delivery Efficiency

With the solution of the governing equations for filler material and HTF, the discharged fluid temperature from a storage tank can be obtained. With the required heat discharge period being given as $t_{ref,discharge}$, an energy delivery effectiveness can be obtained from Eq. (4.1) as discussed before. For convenience of expression, the dimensionless form of the required time period of energy discharge is defined as:

$$\Pi_d = \frac{t_{ref,\,discharge}}{H/U} \tag{4.14}$$

Similarly, a dimensionless form of the time period of energy charge is defined as:

$$\Pi_c = \frac{t_{charge}}{H/U} \tag{4.15}$$

Substituting the dimensionless energy discharge period Π_d into Eq. (4.1), we obtain:

$$\eta = \frac{1}{\Pi_d} \int_0^{\Pi_d} \theta_{f\,(z^*=1,\,t^*)} dt^* \tag{4.16}$$

The energy discharge efficiency will obviously be affected by how much energy is charged into the storage tank. Therefore, it should be noted that η is essentially the function of the following four parameters—Π_c/Π_d, Π_d, τ_r, and H_{CR}. By specifying the dimensionless time period of the discharge process and the mass flow rate, the dimensionless time period of the energy charge can be determined to achieve the objective value of η, which is always desired to approach as close as possible to 1.0.

As has been discussed, a longer energy charging time than energy discharging time is needed in order to achieve an energy delivery effectiveness of approximately 1.0 in a packed-bed system. In addition, the energy storage capacity of a packed-bed tank must be larger than that of an ideal thermal storage tank, as expressed in Eq. (4.2).

4.2.2 Heat Transfer Area S_s and Heat Transfer Coefficient h in Different Types of Storage Systems

As has been already discussed, the governing equations for the temperatures and energy exchange between the primary thermal storage material and the HTF are generally the same for all the thermal storage systems as schematically shown in Fig. 2.5. However, the heat transfer coefficients and the heat transfer area between the primary thermal storage material and HTF for different types of storage systems can be significantly different.

The heat transfer area between rocks and fluid per unit length of tank was denoted as S_s. Therefore, the unit of S_s is in meters. For spherical filler materials, S_s is obtained through the following steps:

(1) The volume of filler material in a unit length Δz of tank is given as $\pi R^2 \Delta z(1-\varepsilon)$.

(2) One sphere of rock has a volume of $V_{sphere} = 4\pi r^3/3$, and therefore, in the length of Δz in the tank, the number of rocks is $\pi R^2 \Delta z(1-\varepsilon)/V_{sphere}$. The total surface area of rocks is then determined to be $\pi R^2 \Delta z(1-\varepsilon) \times 4\pi r^2/V_{sphere}$, which becomes $3\pi R^2(1-\varepsilon)\Delta z/r$ after V_{sphere} is substituted in.

(3) Finally, the heat transfer area of rocks per unit length of tank is:

$$S_s = 3\pi R^2(1-\varepsilon)/r \tag{4.17}$$

The preceding discussion considers the actual volume (assuming ε is known) for solid "spherical particles" in a packed volume. Depending on the packing scheme, the void fraction ε in a packed bed with spheres of a fixed diameter may range from 0.26 to 0.476 [5]. The loosest packaging of spherical rocks in a volume is given by the case where each sphere (of diameter $2r$) is packed into a cube with side lengths of $2r$. The densest packing of spheres causes a void fraction of 0.26, which is due to Kepler's conjecture [6]. Nevertheless, if the packed bed void fraction ε is known, Eq. (4.17) should be used for finding S_s.

The heat transfer coefficient h (W/m$^{2\circ}$C) between the primary thermal storage material (porous media) and HTF can be found in Ref. [7].:

$$h = 0.191\frac{\dot{m}C_f}{\varepsilon\pi R^2}\mathrm{Re}^{-0.278}\mathrm{Pr}^{-2/3} \tag{4.18}$$

where Re is Reynolds number (equal to $4Gr_{char}/\mu_f$) for porous media, as defined by Nellis [7]. The mass flux of fluid through the porous bed is G (equal to $\dot{m}/(\varepsilon\pi R^2)$), and r_{char} is defined as the characteristic radius of the filler material [7], which is equal to $0.25\varepsilon d_r/(1-\varepsilon)$ for a spherical solid filler. Here, d_r is the nominal diameter of a rock, if it is not perfectly spherical.

When the packed particles have irregular shape, such as that of rocks, the convective heat transfer coefficient between the particles and the fluid is a parameter that is difficult to predict accurately. More studies of the convective heat transfer coefficient between packed bed solid materials and fluid are needed for application of thermal energy storage.

4.2.3 Conduction Effects in the Solid Particles or Integrated Solid Region

The model and equations in Sections 4.2.1 and 4.2.2 use the lumped capacitance method to determine the heat transfer inside the filler material. This method actually ignores the resistance to heat conduction inside the filler material. This will result in the calculated energy going into, or coming out from, a filler material being higher than that in the actual physical process. It is known [3] that when the Biot number of the heat transfer of a particle is larger than 0.1, the lumped capacitance assumption will result in increased inaccuracy. In order to correct the lumped capacitance approximation for a spherical "particle" in fluid, Bradshaw et al. [8] and Jefferson [9] proposed to correct the convective heat transfer coefficient between the solid spherical "particle" and the fluid. The modified heat transfer coefficient is then used in the equations for the transient temperature in the particle from the standard lumped capacitance method.

In thermal energy storage systems, there are different types of packed beds of different geometries, as shown in Fig. 2.5. In fact, thermal storage materials could be plates, cylindrical rods immersed in fluid, or a solid core with circular channels inside for HTF. The following sections will present four types of thermal storage packed beds (Fig. 4.3). We use the methodology developed by Bradshaw et al. [8]. The formulas of the effective heat transfer coefficient (Biot number) will be presented for the four types of thermal storage packed beds. Comparison to analytical results will verify their usefulness. Hausen [10] also proposed a similar concept of using corrected heat transfer coefficients in a lumped capacitance method. The equations of corrections to the heat transfer coefficient were also summarized in the book by Schmidt and Willmott [11], which are reexamined in this work. Other works on this issue include those by Razelos and Lazaridis [12], Hughes et al. [13], and Mumma and Marvin [14]. The concluding equations in these references will be compared and evaluated with the results from the current study.

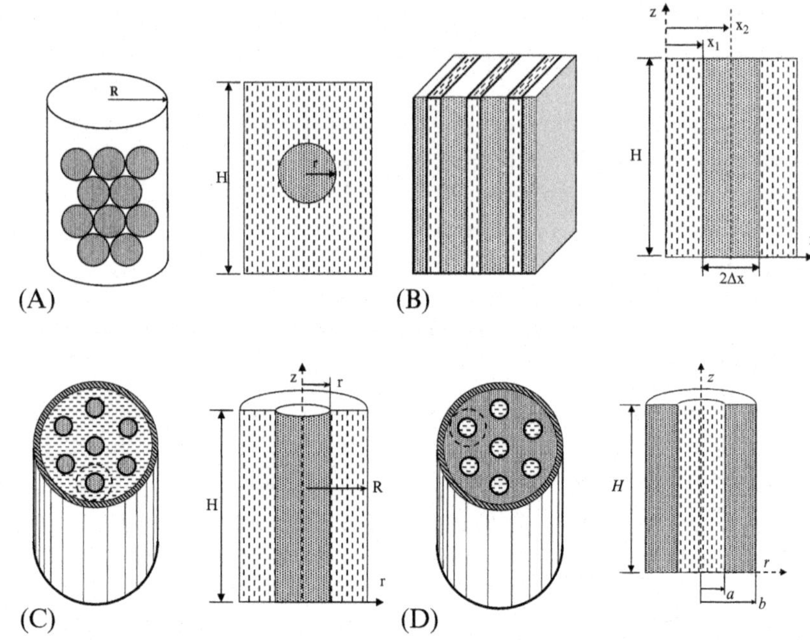

Fig. 4.3 Four typical solid-fluid structural combinations of thermal storage systems (solid thermal storage material; HTF). (A) Packed solid spherical particles with HTF passing around. (B) Packed solid flat plates with HTF passing through the channels in a tank. (C) Packed solid cylinders with HTF passing along in a tank. (D) Solid thermal storage material with HTF tubes passing through in a tank.

The case of a solid sphere is briefly presented here. The readers are referred to the paper by Xu et al. [4] for detailed derivations of all the cases. For thermal storage in a solid material of general geometry, the one-dimensional transient heat transfer governing equations in the fluid and solid are given by Eqs. (4.6) and (4.10). These are rearranged and shown here in dimensional form:

$$\rho_f C_f V_f \left(\frac{\partial T_f}{\partial t} + U \frac{\partial T_f}{\partial z} \right) = h\, S_s \left(T_s - T_f \right) \tag{4.19}$$

$$\rho_s C_s V_s \frac{\partial T_s}{\partial t} = -h\, S_s \left(T_s - T_f \right) \tag{4.20}$$

where V_f is the volume of the fluid per unit length of the thermal storage tank; S_s is the surface area of the solid material in the same control volume per unit length. Eq. (4.19) simulates the heat transfer in the fluid, which has a heat source due to the convection heat transfer between the filler materials and fluid. Eq. (4.20) simulates the heat transfer in the filler material, which has a heat sink term, which is negative but has the exact value as the heat source term in Eq. (4.19). Note that the lumped capacitance assumption for the thermal storage material is assumed to be valid in Eqs. (4.19) and (4.20). The validity of the lumped capacitance assumption is determined by the Biot number, $Bi = hL_p/k_s$, typically less than 0.1, where $L_p = V_s/S_s$.

When the lumped capacitance assumption is not valid, the heat conduction within the solid material has to be included in the formulation, which yields

$$\rho_f C_f V_f \left(\frac{\partial T_f}{\partial t} + U \frac{\partial T_f}{\partial z} \right) = \int_{S_s} -k_s \frac{\partial T_s}{\partial n} dS \tag{4.21}$$

$$\rho_s C_s V_s \frac{\partial T_s}{\partial t} = k_s \nabla^2 T_s \tag{4.22}$$

The following analysis is to approximate the internal heat conduction effect in the thermal storage material, given on the right-hand side of Eq. (4.22), in the form of Newton's cooling relationship. Specifically, this should result in the volume-integrated right-hand side of Eq. (4.22) being expressed as

$$\int_{V_s} k_s \nabla^2 T_s dV = h_{eff} S_s (T_s - T_f) \tag{4.23}$$

Substituting Eq. (4.23) into Eqs. (4.21) and (4.22), we can obtain the lumped capacitance-type of formulation similar to Eqs. (4.19) and (4.20), except the heat transfer coefficient is an effective heat transfer coefficient h_{eff}. In fact, the effective heat transfer coefficient h_{eff} is a corrected value from the heat transfer coefficient h. It is necessary to point out that the effective heat transfer coefficient h_{eff} will strongly depend on the geometry of the solid filler materials. The following analysis takes spherical filler material as an example to elucidate the methodology. Following the method, we will present similar solutions for the other three geometries of thermal storage materials shown in Fig. 4.3.

For the purpose of extracting the effective heat transfer coefficient as explained previously, it is sufficient to consider a simple case where the fluid temperature does not vary with time and therefore the heat transfer mainly causes the fluid temperature to change in the z direction only. This simplification allows us to focus on the heat transfer between the fluid and filler materials. Therefore, the energy balance equations from the lumped capacitance method for the fluid and thermal storage material are

$$\rho_f C_f V_f U \frac{\partial T_f}{\partial z} = h_{eff}\, S_s (T_s - T_f) \tag{4.24}$$

$$\rho_s C_s V_s U \frac{\partial T_s}{\partial t} = -h_{eff}\, S_s (T_s - T_f) \tag{4.25}$$

The initial conditions are

$$t = 0, \quad T_f = T_s = 0 \tag{4.26}$$

The boundary conditions are

$$z = 0,\ T_f = TI(t) \tag{4.27}$$

where $TI(t)$ is the fluid inlet temperature. We further assume that the solid material attains the final equilibrium temperature, M (independent of z), at the end of the charging process, in time τ. The fluid temperature at $z = L$ is $TO(t)$. Define the finite Laplace transform as

$$\Lambda_f(z, p) = \int_0^{\tau} T_f(z, t) e^{-pt} dt \tag{4.28}$$

and

$$\Lambda_s(z, p) = \int_0^{\tau} T_s(z, t) e^{-pt} dt \tag{4.29}$$

Applying these finite Laplace transforms to Eqs. (4.24) and (4.25) and eliminating $\Lambda_s(z, p)$, we obtain

$$\frac{d\Lambda_f}{dz} + \frac{\beta p}{p+\gamma} \Lambda_f + \frac{\beta M_e^{-p\tau}}{p+\gamma} = 0 \tag{4.30}$$

where M is the equilibrium temperature (constant) between the solid and the HTF,

$$\beta = \frac{hs_s}{\rho_f\, C_f \varepsilon\, U} \tag{4.31}$$

$$\gamma = \frac{hs_s}{\rho_s\, C_s} \tag{4.32}$$

The transformed boundary condition, Eq. (4.27), is

$$z = 0, \quad f(p) = \int_0^{\tau} TI(t) e^{-pt} dt \tag{4.33}$$

The solution to Eq. (4.30) subject to boundary conditions as in Eq. (4.33) is

$$\Lambda_f(z,p) = -\frac{M}{p} e^{-p\tau} + \left[f(p) + \frac{M}{p} e^{-p\tau} \right] e^{-\frac{\beta p z}{p+\gamma}} \tag{4.34}$$

It is interesting to examine the following formula for weighted average time with weighting function $g(t)$, which was introduced by Bradshaw et al. [8]:

$$\langle t \rangle = \frac{\int_0^{\tau} t g(t) dt}{\int_0^{\tau} g(t) dt} = \lim_{p \to 0} \frac{\int_0^{\tau} t g(t) e^{-pt} dt}{\int_0^{\tau} g(t) e^{-pt} dt} = -\lim_{p \to 0} \frac{\frac{d}{dp} \int_0^{\tau} g(t) e^{-pt} dt}{\int_0^{\tau} g(t) e^{-pt} dt} \tag{4.35}$$

The weighted average time can therefore be calculated by the following equation:

$$\langle t \rangle = -\lim_{p \to 0} \frac{\frac{d}{dp} \Gamma(p)}{\Gamma(p)} \tag{4.36}$$

where $\Gamma(p)$ is the finite Laplace transform of function $g(t)$.

We will now form a function, $\Delta\Lambda_f(p) = \Lambda_f(0,p) - \Lambda_f(L,p)$, which is the difference of the finite Laplace transform of the fluid temperature at the inlet and outlet. Substituting the function $\Delta\Lambda_f(p)$ as $\Gamma(p)$ into Eq. (4.36), the weighted average time formula for the lumped capacitance case is obtained as

$$\langle t \rangle = \frac{\rho_s C_s}{h_{eff} S_s} + \tau + \frac{\rho_s C_s}{2U \rho_f C_f \varepsilon} - \frac{f(0)}{M} \tag{4.37}$$

This is the functional form for the weighted average time.

We now switch our attention to the conduction effects within the solid. The heat transfer within the solid body is modeled by the heat diffusion equation in the appropriate coordinate system. Following Bradshaw et al.'s [8] method, we obtain the analytical formulas for the weighted average time. By comparing these formulas to Eq. (4.37), we can extract the effective heat transfer coefficient h_{eff}. Therefore, the governing equations are

$$\rho_f C_f V_f U \frac{\partial T_f}{\partial z} = -k_s S_s \left(\frac{\partial T_s}{\partial r} \right)_{r=r_s} \tag{4.38}$$

$$\rho_s C_s \frac{\partial T_s}{\partial z} = k_s \frac{1}{r^2} \frac{\partial}{\partial r} \left(r^2 \frac{\partial T_s}{\partial r} \right) \tag{4.39}$$

The initial and boundary conditions for T_f remain the same as in Eqs. (4.26) and (4.27). The initial and boundary conditions for T_s are:

$$T_s(r, 0) = 0 \tag{4.40}$$

$$r = 0, \; \frac{\partial T_s}{\partial r} = 0 \tag{4.41}$$

$$r = r_s, \; k_s \frac{\partial T_s}{\partial r} = h(T_f - T_s) \tag{4.42}$$

Taking the finite Laplace transform, we have

$$\rho_f C_f V_f U \frac{d\Lambda_f}{dz} + k_s S_s \left(\frac{\partial \Lambda_s}{\partial r} \right)_{r=r_s} = 0 \tag{4.43}$$

$$\rho_s C_s (M e^{-p\tau} + p\Lambda_s) = k_s \frac{1}{r^2} \frac{\partial}{\partial r} \left(r^2 \frac{\partial T_s}{\partial r} \right) \tag{4.44}$$

$$r = 0, \; \frac{\partial \Lambda_s}{\partial r} = 0 \tag{4.45}$$

$$r = r_s, \; k_s \frac{\partial \Lambda_s}{\partial r} = h(\Lambda_f - \Lambda_s) \tag{4.46}$$

The transformed temperature of the sphere can be solved as

$$\Lambda_s(r, p) = A(r, p) + B(r, p)\, \Lambda_f(z, p) \tag{4.47}$$

where

$$A(r,p) = -\frac{e^{-p\tau} M}{p} + \frac{\sin\mathrm{h}\left[r\sqrt{\frac{p S_s}{V_f k_s (1-\varepsilon)}} \right] \left[\frac{e^{-p\tau} M k_s}{p r_s} + \frac{e^{-p\tau} M r_s \left(-\frac{k_s}{r_s^2} + \frac{h}{r_s} \right)}{p} \right]}{r \left[\frac{k_s \sqrt{\frac{p S_s}{V_f k_s (1-\varepsilon)}} \cos\mathrm{h}\left[r_s \sqrt{\frac{p S_s}{V_f k_s (1-\varepsilon)}} \right]}{r_s} + \left(-\frac{k_s}{r_s^2} + \frac{h}{r_s} \right) \sin\mathrm{h}\left[r_s \sqrt{\frac{p S_s}{V_f k_s (1-\varepsilon)}} \right] \right]} \tag{4.48}$$

$$B(r,p) = \frac{\sinh\left[r\sqrt{\frac{p\,S_s}{V_f k_s(1-\varepsilon)}}\right]h}{r\left[\frac{k_s\sqrt{\frac{p\,S_s}{V_f k_s(1-\varepsilon)}}\cosh\left[r_s\sqrt{\frac{p\,S_s}{V_f k_s(1-\varepsilon)}}\right]}{r_s} + \left(-\frac{k_s}{r_s^2}+\frac{h}{r_s}\right)\sinh\left[r_s\sqrt{\frac{p\,S_s}{V_f k_s(1-\varepsilon)}}\right]\right]} \tag{4.49}$$

Equation (4.43) can then be solved for $\Lambda_f(r,p)$,

$$\Lambda_f(z,p) = \frac{A(r_s,p)}{B(r_s,p)}\left(e^{\frac{(B(r_s,p)-1)hS_s}{\rho_f C_f V_f U}z} - 1\right) + f(0)e^{\frac{(B(r_s,p)-1)hS_s}{\rho_f C_f V_f U}z} \tag{4.50}$$

We now follow the same procedure as in the lumped capacitance case by defining $\Delta\Lambda_f(p) = \Lambda_f(0,p) - \Lambda_f(L,p)$. The weighted average time is obtained for the solid sphere, which has internal resistance

$$\begin{aligned}\langle t\rangle &= \frac{\rho_s C_s}{h\,S_s} + \frac{r_s\rho_s C_s}{5\,k_s\,S_s} + \tau + \frac{\rho_s C_s}{2U\rho_f C_f\varepsilon} - \frac{f(0)}{M} \\ &= \frac{\rho_s C_s}{h_{eff}S_s} + \tau + \frac{\rho_s C_s}{2U\rho_f C_f\varepsilon} - \frac{f(0)}{M}\end{aligned} \tag{4.51}$$

In Eq. (4.51) we use the two terms $\rho_s C_s/h\ S_s$ and $r_s\rho_s C_s/5\ k_s\ S_s$ as one term and introduce an effective heat transfer coefficient h_{eff}. This makes the structure of Eq. (4.51) similar to the equation from the lumped capacitance case given by Eq. (4.37). Therefore, for the sphere solid thermal storage material, the relationship of the effective heat transfer coefficient h_{eff} and the actual heat transfer coefficient h is obtained:

$$h_{eff-sp} = \frac{1}{\frac{1}{h} + \frac{r_s}{5k_s}} \tag{4.52}$$

When h_{eff-sp} is used to replace h in the lumped capacitance method, we expect the solution will approach the results from a precise analytical solution. The demonstration of this is given in Section 4.4.1. The current result for the effective heat transfer coefficient of a sphere matches the solution published by Jefferson [9]. For three other cases that have not been discussed by Bradshaw et al. [8] and Jefferson [9], the same methodology can be applied. The resulting effective heat transfer coefficient formulae are tabulated in Table 4.1.

Table 4.1 The effective heat transfer coefficients of solid thermal storage materials of different structures

	Characteristic length for lumped capacitance Biot number	Effective heat transfer coefficient h_{eff}	Effective lumped capacitance Biot number Bi_{eff}	Lumped capacitance Biot number Bi_{LC}
Sphere	$\frac{R}{3}$	$\frac{1}{\frac{1}{h}+\frac{R}{5k_r}}$	$\frac{Bi_{LC}}{1+\frac{3}{5}Bi_{LC}}$	$\frac{h}{k_r}\frac{R}{3}$
Plate	$\Delta x = x_2 - x_1$ See Fig. 4.3B for definition of x_1, x_2	$\frac{1}{\frac{1}{h}+\frac{(x_2-x_1)}{3k_r}}$	$\frac{Bi_{LC}}{1+\frac{1}{3}Bi_{LC}}$	$\frac{h}{k_r}\Delta x$
Cylinder	$\frac{R}{2}$	$\frac{1}{\frac{1}{h}+\frac{R}{4k_r}}$	$\frac{Bi_{LC}}{1+\frac{1}{2}Bi_{LC}}$	$\frac{h}{k_r}\frac{R}{2}$
Tube with fluid inside and outside insulated	$\frac{b^2-a^2}{2a}$ See Fig. 4.3D for definition of a, b; $\eta = b/a$	$\frac{1}{\frac{1}{h}+\frac{1}{k_r}\frac{a^3(4b^2-a^2)+ab^4(4Ln[b/a]-3)}{4(b^2-a^2)^2}}$	$\frac{Bi_{LC}}{1+Bi_{LC}\frac{\eta^4[4Ln(\eta)-3]+4\eta^2-1}{2(\eta^2-1)^3}}$	$\frac{h}{k_r}\frac{b^2-a^2}{2a}$

The effectiveness of the effective heat transfer coefficient can be examined by considering the heat conduction problem of a sphere subject to convective heat cooling/heating. Analytical solutions are available in textbooks [15,16]. For all the discussed solid bodies of different shapes in this study, the equations for dimensionless temperatures and heat storage/discharge are listed in Table 4.2. The definitions of the general *Bi* used in analytical solutions for different cases are also given in Table 4.2.

Fig. 4.4 shows the dimensionless heat absorbed/released in a solid body versus time determined from the lumped capacitance method, corrected lumped capacitance method, and analytical method for a sphere. At a small Biot number, 0.1, the curves of the dimensionless energy storage from all three methods agree very well, which verifies that the lumped capacitance method is valid at small Biot numbers. At a Biot number of 1.0, the corrected lumped capacitance method agrees with the analytical solution very well, while the curve from the lumped capacitance method has a significant discrepancy with the analytical solution. When the Biot number is 10.0 or 100, the discrepancy between the results from the corrected lumped capacitance method and the analytical method increases slightly, but is still acceptable. However, the lumped capacitance method predicts a very different value of energy storage and is thus unacceptable.

For other cases of fluid-solid configuration in Table 4.2, comparison of the energy storage in the solid material based on computation from the lumped capacitance method, corrected lumped capacitance method, and accurate analytical solution was detailed in the work by Xu et al. [4]. As shown in Fig. 4.4, it has been proven that the corrected lumped capacitance method can predict the energy storage with almost the same accuracy as that of the analytical method, while the original lumped capacitance method cannot predict with acceptable accuracy.

4.2.4 Numerical Solution for the One-Dimensional Model for Dual-Media Sensible Thermal Storage

A number of analyses and solutions to the heat transfer governing equations of a working fluid flowing through a filler packed bed have been presented in the past (Schumann [17]; Shitzer & Levy [18]; McMahan [19]; Beasley and Clark [20]; Zarty & Juddaimi [21]). As the pioneering work, Schumann [17] presented a set of equations governing the energy conservation of fluid flow through porous media. Schumann's equations have been widely adopted in the analysis of thermocline heat storage utilizing solid filler

Table 4.2 Dimensionless temperature and energy in a solid body from analytical solution [15]

Solid body	Solution
Sphere	$\theta_{sp}(r^*, t^*) = \sum_{m=1}^{\infty} \left[\frac{4\left(\sin(\overline{\beta}_m) - \overline{\beta}_m \cos(\overline{\beta}_m)\right)}{2\overline{\beta}_m - \sin(\overline{\beta}_m)} \right] \frac{\sin(\overline{\beta}_m r^*)}{\overline{\beta}_m r^*} e^{-\overline{\beta}_m^2 t^*}$
	$1 - \overline{\beta}_m \cdot \cot(\overline{\beta}_m) = Bi;\ Bi = hR/k_r$
	$Q_{sp}^*(t^*) = 1 - e^{-\overline{\beta}_m^2 t^*} \sum_{m=1}^{\infty} \left[\frac{4\left[\sin(\overline{\beta}_m) - \overline{\beta}_m \cos(\overline{\beta}_m)\right]}{2\overline{\beta}_m - \sin(2\overline{\beta}_m)} \right] \frac{4\pi\left(\sin(\overline{\beta}_m) - \overline{\beta}_m \cos(\overline{\beta}_m)\right)}{\overline{\beta}_m^3}$
Plate	$\theta_p(x^*, t^*) = \sum_{m=1}^{\infty} \frac{2\left(\overline{\beta}_m^2 + Bi^2\right)}{\overline{\beta}_m^2 + Bi^2 + Bi} \frac{\sin(\overline{\beta}_m)}{\overline{\beta}_m} \cos\left[\overline{\beta}_m(1 - x^*)\right] e^{-\overline{\beta}_m^2 t^*}$
	$\overline{\beta}_m \cdot \tan(\overline{\beta}_m) = Bi;\ Bi = h\Delta x/k_r$
	$Q_p^*(t^*) = 1 - e^{-\overline{\beta}_m^2 t^*} \sum_{m=1}^{\infty} \left[\frac{2\left(\overline{\beta}_m^2 + Bi^2\right)}{\overline{\beta}_m^2 + Bi^2 + Bi} \left(\frac{\sin(\overline{\beta}_m)}{\overline{\beta}_m} \right)^2 \right]$
Cylinder	$\theta_{cy-1}(r^*, t^*) = \sum_{m=1}^{\infty} \left[\frac{2\overline{\beta}_m^2}{J_0^2(2\overline{\beta}_m)\left(Bi^2 + \overline{\beta}_m^2\right)} \frac{J_1(2\overline{\beta}_m)}{2\overline{\beta}_m} \right] J_0(2r^*\overline{\beta}_m) e^{-4\overline{\beta}_m^2 t^*}$
	$2\overline{\beta}_m \cdot J_1(2\overline{\beta}_m R) = Bi \cdot J_0(2\overline{\beta}_m R);\ Bi = hR/k_r$
	$Q_{cy-1}^*(t^*) = \int_0^1 2(1 - \theta_2) r^* dr^* = 2\left(1 - \sum_{m=1}^{\infty} \left[\frac{2\overline{\beta}_m^2}{J_0^2(2\overline{\beta}_m)\left(Bi^2 + \overline{\beta}_m^2\right)} \left(\frac{J_1(2\overline{\beta}_m)}{2\overline{\beta}_m} \right)^2 \right] e^{-4\overline{\beta}_m^2 t^*} \right)$

Tube	$\theta_{cy-2}(r^*, t^*) = \sum_{m=1}^{\infty} e^{-\overline{\beta}_m^2 t^*} \left[\frac{\pi^2}{2} \frac{\overline{\beta}_m^2 G(\overline{\beta}_m)}{G(\overline{\beta}_m) - \left(Bi^2 + \overline{\beta}_m^2\right) J_1^2(\eta \overline{\beta}_m)} \right] H(\overline{\beta}_m) R(r^*)$ $R(r^*) = J_1(\overline{\beta}_m \eta) Y_0(\overline{\beta}_m r^*) - J_0(\overline{\beta}_m r^*) Y_1(\overline{\beta}_m \eta)$; $Bi = ha/k_r$ $H(\overline{\beta}_m) = \int_1^{\eta} r^* R(r^*) dr^*$ $G(\overline{\beta}_m) = \left(\overline{\beta}_m J_1(\overline{\beta}_m) + Bi J_0(\overline{\beta}_m)\right)^2$ where the eigenvalues $\overline{\beta}_m$ are determined from the following equation: $[\overline{\beta}_m J_1(\overline{\beta}_m) + Bi J_0(\overline{\beta}_m)] Y_1(\overline{\beta}_m \eta) = [\overline{\beta}_m Y_1(\overline{\beta}_m) + Bi Y_0(\overline{\beta}_m)] J_1(\overline{\beta}_m \eta)$ $Q^*_{cy-2}(t^*) = \frac{2}{(\eta^2 - 1)^2} \int_1^{\eta} (1 - \theta_{cy-2}) r^* dr^* = 1 - \frac{1}{(\eta^2 - 1)^2} \sum_{m=1}^{\infty} e^{-\overline{\beta}_m^2 t^*} \left[\frac{\pi^2 \overline{\beta}_m^2 G(\overline{\beta}_m)}{G(\overline{\beta}_m) - \left(Bi^2 + \overline{\beta}_m^2\right) J_1^2(\eta \overline{\beta}_m)} \right] H^2(\overline{\beta}_m)$

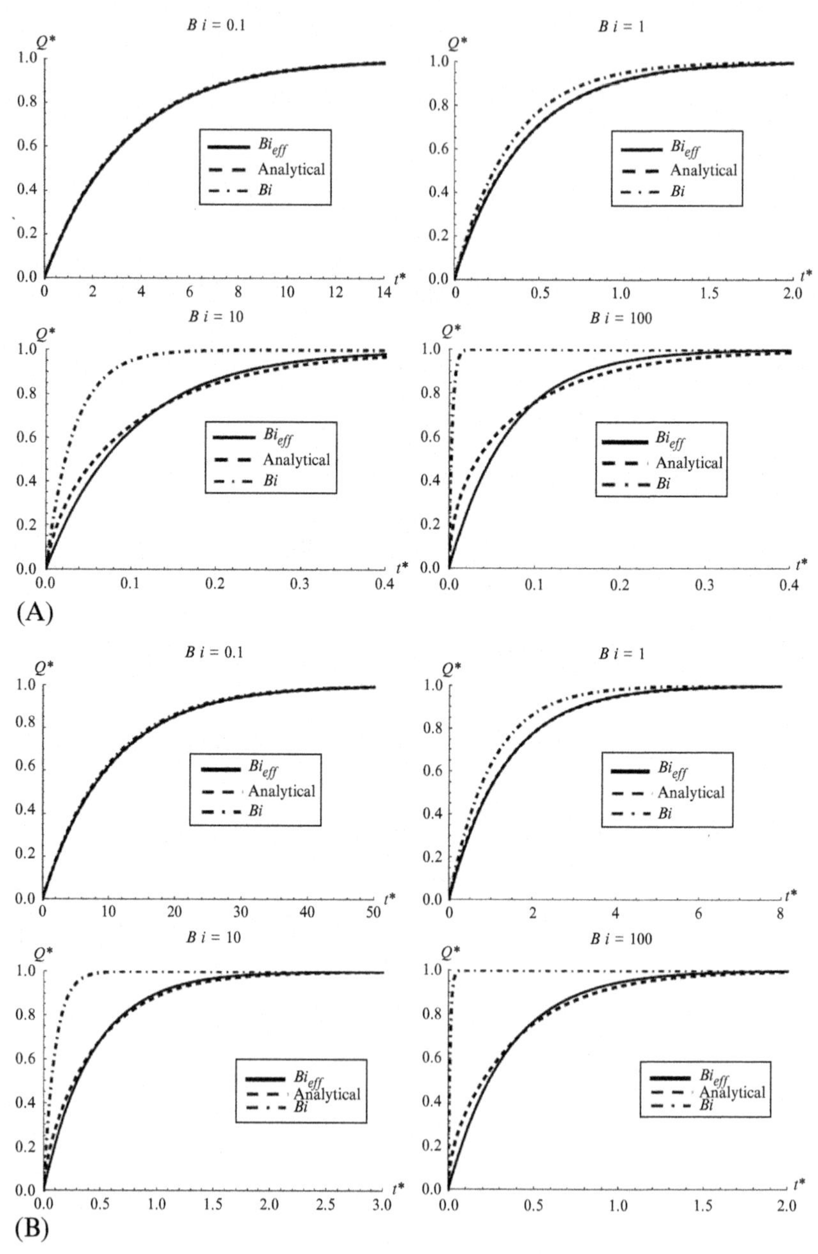

Fig. 4.4 Comparison of results from analytical method (Q^* given in Table 4.2), lumped capacitance method (with Bi used and $Q^* = 1 - e^{-Bi\,t^*}$), and corrected lumped capacitance method (with Bi_{eff} used and $Q^* = 1 - e^{-Bi_{eff}\,t^*}$) at different Biot numbers. The definition of dimensionless time is $t^* = \frac{k_s}{\rho_s C_s} \frac{t}{(V_s/S_s)^2}$. (A) Spheres; (B) plate;

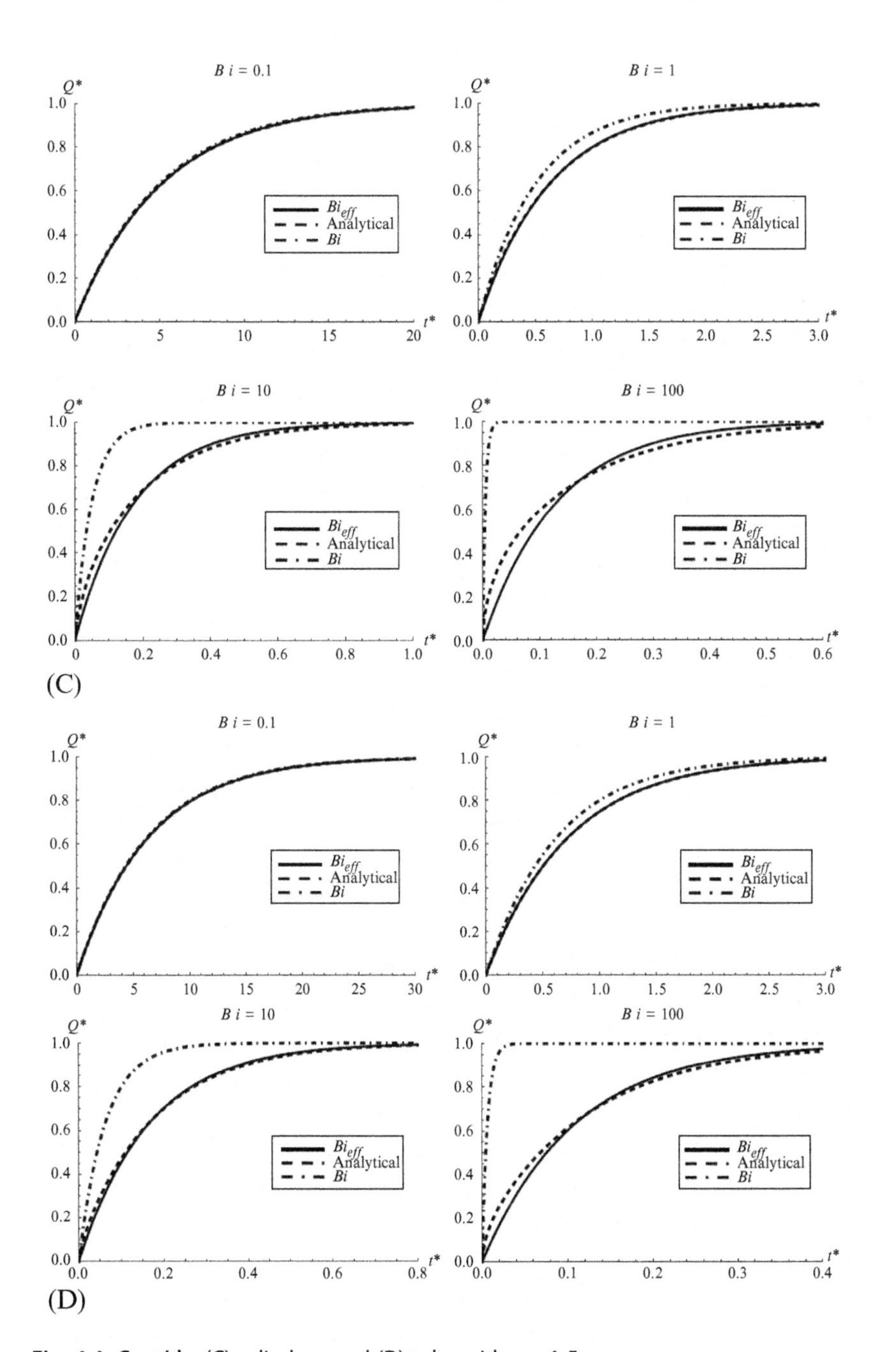

Fig. 4.4, Cont'd (C) cylinders; and (D) tube with $\eta = 1.5$.

material inside a tank. His analysis and solutions were for the special case where there is a fixed fluid temperature at the inlet to the storage system. In most solar thermal storage applications, this may not be the actual situation. To overcome this limitation, Shitzer and Levy [18] employed Duhamel's theorem on the basis of Schumann's solution to consider a transient inlet fluid temperature to the storage system. The analyses of Schumann and Shitzer and Levy, however, still carry with them some limitations. Their method does not consider a nonuniform initial temperature distribution. For a heat storage system, particularly in a solar thermal power plant, heat charge and discharge are cycled daily. The initial temperature field of a heat charge process is dictated by the most recently completed heat discharge process, and vice versa. Therefore, nonuniform and nonlinear temperature distribution is typical for both charge and discharge processes. To consider a nonuniform initial temperature distribution and varying fluid temperature at the inlet in a heat storage system, numerical methods have been deployed by researchers in the past.

To avoid the long mathematical analysis necessary in analytical solutions, numerical methods used to solve the Schumann equations were discussed in the literature by McMahan [19,22] and Pacheco et al. [23], and demonstrated in the TRNSYS software developed by Kolb and Hassani [24]. Based on the regular finite-difference method, McMahan gave both explicit and implicit discretized equations for the Schumann equations. Whereas the explicit solution method had serious solution stability issues, the implicit solution method encountered an additional computational overhead, thus requiring a dramatic amount of computation time. The solution for the complete power plant with thermocline storage provided by the TRNSYS model in the work of Kolb and Hassani [24] cites the short time step requirement for the differential equations of the thermocline as one major source of computer time consumption. To overcome the problems encountered in the explicit and implicit method, McMahan et al. [22] also proposed an infinite-NTU method. This model, however, is limited to the case in which the heat transfer of the fluid compared to the heat storage in fluid is extremely large.

The present study has approached the governing equations using a different numerical method [25]. The governing equations have been reduced to dimensionless forms, which allow for a universal application of the solution. The dimensionless hyperbolic type equations are solved numerically by the method of characteristics. This numerical method overcomes the numerical difficulties encountered in McMahan's work—explicit, implicit, and the restriction on infinite-NTU method (McMahan [19] and McMahan et al. [22]). The current model gives a direct solution to the discretized equations (with no iterative computation needed) and completely eliminates any computational overhead. A grid-independent solution is obtained at a small number of nodes. The method of characteristics and the present numerical solution has proven to be a fast, efficient, and accurate algorithm for the Schumann equations.

The nondimensional energy balance equations for the HTF and rocks can be solved numerically along the characteristics [26–28]. Eq. (4.8) can be reduced along the characteristic $t^* = z^*$ so that:

$$\frac{D\theta_f}{Dt^*} = \frac{1}{\tau_r}\left(\theta_s - \theta_f\right) \tag{4.53}$$

Separating and integrating along the characteristic, the equation becomes:

$$\int d\theta_f = \int \frac{1}{\tau_r}\left(\theta_s - \theta_f\right) dt^* \tag{4.54}$$

Similarly, Eq. (4.11) for the energy balance of rocks is reposed along the characteristic $z^* =$ constant so that:

$$\frac{d\theta_s}{dt^*} = -\frac{H_{CR}}{\tau_r}\left(\theta_s - \theta_f\right) \tag{4.55}$$

The solution for Eq. (4.55) is very similar to that for Eq. (4.53) but with the additional factor of H_{CR}. The term H_{CR} is simply a ratio of fluid heat capacitance to rock heat capacitance. Therefore, the equation for the solution of θ_s will react with a dampened speed when compared to θ_f, as the filler material must have the capacity to store the energy being delivered to it, or vice versa. Finally, separating and integrating along the characteristic for Eq. (4.55) results in:

$$\int d\theta_s = -\int \frac{H_{CR}}{\tau_r}\left(\theta_s - \theta_f\right) dt^* \tag{4.56}$$

There are now two characteristic equations bound to intersections of time and space. A discretized grid of points, laid over the time-space

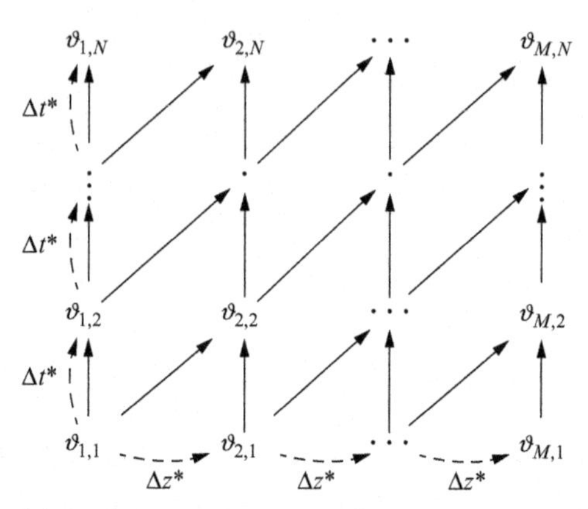

Fig. 4.5 Diagram of the solution matrix arising from the method of characteristics.

dimensions, will have nodes at these intersecting points. A diagram of these points in a matrix is shown in Fig. 4.5. In space, there are $i=1, 2, \ldots, M$ nodes broken up into step sizes of Δz^* to span all of z^*. Similarly, in time, there are $j=1, 2, \ldots, N$ nodes broken up into time-steps of Δt^* to span all of t^*. Looking at a grid of the ϑ nodes, a clear picture of the solution can arise. To demonstrate a calculation of the solution we can look at a specific point in time, along z^*, where there are two points, $\vartheta_{1,1}$ and $\vartheta_{2,1}$. These two points are the starting points of their respective characteristic waves described by Eqs. (4.53) and (4.56). After the time Δt^* there is a third point $\vartheta_{2,2}$ which has been reached by both wave equations. Therefore, Eq. (4.54) can be integrated numerically as:

$$\int_{\vartheta_{1,1}}^{\vartheta_{2,2}} d\theta_f = \int_{\vartheta_{1,1}}^{\vartheta_{2,2}} \frac{1}{\tau_r}\left(\theta_s - \theta_f\right) dt^* \tag{4.57}$$

The numerical integration of the right-hand side is performed via the trapezoidal rule and the solution is:

$$\theta_{f_{2,2}} - \theta_{f_{1,1}} = \frac{1}{\tau_r}\left(\frac{\theta_{s_{2,2}} + \theta_{s_{1,1}}}{2} - \frac{\theta_{f_{2,2}} + \theta_{f_{1,1}}}{2}\right)\Delta t^* \tag{4.58}$$

where $\theta_{f_{1,1}}$ is the value of θ_f at $\vartheta_{1,1}$, and $\theta_{f_{2,2}}$ is the value of θ_f at $\vartheta_{2,2}$, and similarly so for θ_r.

The integration for Eq. (4.36) along $z^* = constant$ constant is:

$$\int_{\vartheta_{2,1}}^{\vartheta_{2,2}} d\theta_s = \int_{\vartheta_{2,1}}^{\vartheta_{2,2}} -\frac{H_{CR}}{\tau_r}\left(\theta_s - \theta_f\right) dt^* \tag{4.59}$$

The numerical integration of the right-hand side is also performed via the trapezoidal rule and the solution is:

$$\theta_{s_{2,2}} - \theta_{s_{2,1}} = -\frac{H_{CR}}{\tau_r}\left(\frac{\theta_{s_{2,2}} + \theta_{s_{2,1}}}{2} - \frac{\theta_{f_{2,2}} + \theta_{f_{2,1}}}{2}\right)\Delta t^* \tag{4.60}$$

Eqs. (4.58) and (4.60) can be reposed as a system of algebraic equations for two unknowns, $\theta_{f_{2,2}}$ and $\theta_{r_{2,2}}$, while θ_f and θ_r at grid points $\vartheta_{1,1}$ and $\vartheta_{2,1}$ are known.

$$\begin{bmatrix} 1 + \dfrac{\Delta t^*}{2\tau_r} & -\dfrac{\Delta t^*}{2\tau_r} \\ -\dfrac{H_{CR}\Delta t^*}{2\tau_r} & 1 + \dfrac{H_{CR}\Delta t^*}{2\tau_r} \end{bmatrix} \begin{bmatrix} \theta_{f_{2,2}} \\ \theta_{s_{2,2}} \end{bmatrix} = \begin{bmatrix} \theta_{f_{1,1}}\left(1 - \dfrac{\Delta t^*}{2\tau_r}\right) + \theta_{r_{1,1}}\dfrac{\Delta t^*}{2\tau_r} \\ \theta_{f_{2,1}}\dfrac{H_{CR}\Delta t^*}{2\tau_r} + \theta_{r_{2,1}}\left(1 - \dfrac{H_{CR}\Delta t^*}{2\tau_r}\right) \end{bmatrix} \tag{4.61}$$

Cramer's rule [28] can be applied to obtain the solution efficiently. It is important to note that all coefficients/terms in Eq. (4.61) are independent of z^*, t^*, θ_f, and θ_s, thus they can be evaluated once for all. Therefore, the numerical computation takes a minimum of computing time, and is much more efficient than the method applied in references [19,22].

From the grid matrix in Fig. 4.5 it is seen that the temperatures of the rock and fluid at grids $\vartheta_{i,1}$ are the initial conditions. The temperatures of the fluid and rock at grid $\vartheta_{1,1}$ are the inlet conditions, which vary with time. The inlet temperature for the fluid versus time is given. The rock temperature (as a function of time) at the inlet can be easily obtained using Eq. (4.11), for which the inlet fluid temperature is known. Now, as the conditions at $\vartheta_{1,1}$, $\vartheta_{1,2}$, and $\vartheta_{2,1}$ are known, the temperatures of the rocks and fluid at $\vartheta_{2,2}$ will be easily calculated from Eq. (4.61).

Extending the preceding sample calculation to all points in the ϑ grid of time and space will give the entire matrix of solutions in time and space for both the rocks and fluid. While the march of Δz^* steps is limited to $z^* = 1$ the march of time Δt^* has no limitation.

The previous numerical integrations used the trapezoidal rule; the error of such an implementation is not straightforwardly analyzed but the formal accuracy is on the order of $O(\Delta t^* 2)$ for functions [28] such as those solved in this study.

4.2.5 Computer Code for the One-Dimensional Computation Analysis

A MATLAB computer code has been developed to accomplish the simulation and analysis for sensible thermal energy storage. The code can be used to compute cases with any of the listed configurations of solid packed bed with HTFs flowing through, as shown in Fig. 4.3.

A brief introduction to the structure of the computer code and the MATLAB code is attached as an appendix in this chapter.

4.2.6 Numerical Results for the Temperature Variation in Packed Bed Sensible Energy Storage

The first analysis of the storage system was done on a single tank configuration of a chosen geometry, using a filler and fluid with given thermodynamic properties. The advantage of having the governing equations reduced to their dimensionless form is that by finding the values of two dimensionless parameters (τ_r and H_{CR}) all the necessary information about the problem is known. The properties of the fluid and filler rocks, as well as the tank dimensions, which determines τ_r and H_{CR} for the example problem, are summarized in Table 4.3.

The numerical computation started from a discharge process assuming initial conditions of an ideally charged tank with the fluid and rocks both having the same high temperature throughout the entire tank, i.e., $\theta_f = \theta_s = 1$. After the heat discharge, the temperature distribution in the tank is taken as the initial condition of the following charge process. The discharge and charge time were each set to 4 h. The fluid mass flow rate was determined such that an empty (no filler) tank was sure to be filled by the fluid in 4 h. It was found that, with the current configuration, after five discharge and charge cycles the results of all subsequent discharge processes were identical—likewise for the charge processes. It is therefore assumed that the solution is then independent of the first initial condition. The data presented in the following portions of this section are the results from the cyclic discharge and charge processes after five cycles.

Shown in Fig. 4.6 are the temperature profiles in the tank during a discharge process, in which cold fluid enters the tank from the bottom of the tank. The location of $z^* = 0$ is at the bottom of a tank for a discharge process. The temperature profile evolves as discharging proceeds, showing the heat wave propagation and the high temperature fluid moving out of the storage tank. The fluid temperature at the exit ($z^* = 1$) of the tank gradually

Table 4.3 Dimensions and parameters of a thermocline tank [25]

ε	τ_r	H_{CR}	H	R	t
0.25	0.0152	0.3051	14.6 m	7.3 m	4 h

Fluid (Therminol VP-1) properties:

$T_H = 395°C$	$T_L = 310°C$	$\rho_f = 753.75 kg/m^3$	$C_f = 2474.5 J/(kg K)$	$k_f = 0.086 W/(m K)$;	$\dot{m} = 128.74 kg/s$	$\mu_f = 1.8 \times 10^{-4} Pa s$

Filler material (granite rocks) properties:

$\rho_s = 2630 kg/m^3$	$C_s = 775 J/(kg K)$	$k_s = 2.8 W/(m K)$	$d_r = 0.04 m$

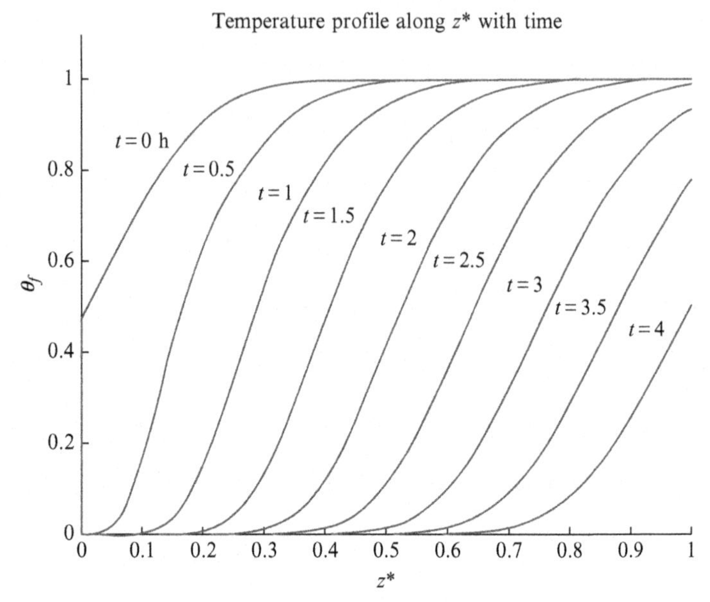

Fig. 4.6 Dimensionless fluid temperature profile in the tank for every 0.5 h.

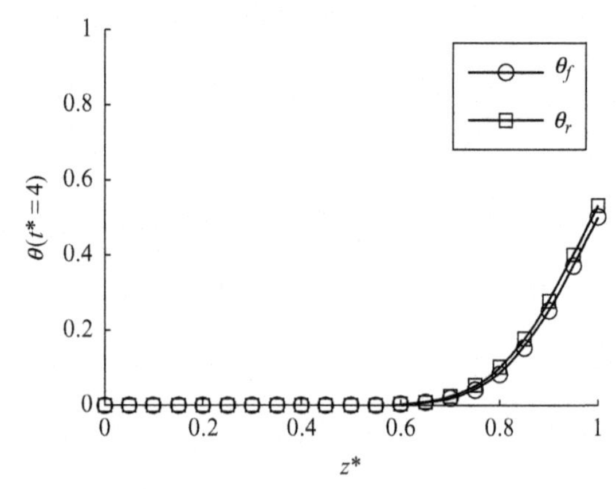

Fig. 4.7 Dimensionless temperature distribution in the tank after time $t^*=4$ of discharge (here θ_r is used to denote θ_s, as rocks are used as the storage material in the example).

decreases after 3 h of discharge. At the end of the discharge process, the temperature distribution along the tank is shown in Fig. 4.7. It is seen that the fluid and rock temperatures, θ_f and θ_s respectively (θ_s is denoted by θ_r when the filler material is rock), in the region with z^* below 0.7 are almost zero, which means that the heat in the rocks in this region has been completely extracted by the passing fluid. In the region from $z^*=0.7$ to $z^*=1.0$ the

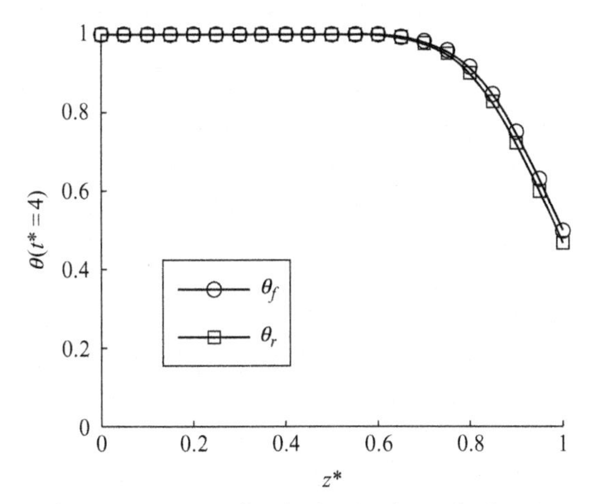

Fig. 4.8 Dimensionless temperature distribution in the tank after time $t^*=4$ of charge (here θ_r is used to denote θ_s, as rocks are used as the storage material in the example).

temperature of the fluid and rock gradually becomes higher, which indicates that some heat has remained in the tank.

A heat charge process has a similar heat wave propagation scenario. The temperature for the filler and fluid along the flow direction is shown in Fig. 4.8 after a 4 h charging process. During a charge process, fluid flows into the tank from the top, where z^* is set as zero. It is seen that for the bottom region (z^* from 0.7 to 1.0) the temperatures of the fluid and rocks decrease significantly. A slight temperature difference between the HTF and the rocks also exists in this region.

The next plots of interest are the variation of θ_f at $z^*=1$ as dimensionless time progresses for a charging or discharging process. Fig. 4.9 shows the behavior of θ_f at the outlet during both charge and discharge cycles. For the charge cycle, θ_f begins to increase when all of the initially cold fluid has been ejected from the thermocline tank. For the present thermocline tank, the fluid that first entered the tank at the start of the cycle has moved completely through the tank at $t^*=1$, which also indicates that the initially existing cold fluid of the tank has been ejected from the tank. Similarly, during the discharge cycle, after the initially existing hot fluid in the tank has been ejected, the cold fluid that first entered the tank from the bottom at the start of the cycle has moved completely through the tank at $t^*=1$. At $t^*=2.5$ or $t=2.5\,\text{h}$, the fluid temperature θ_f starts to drop. This is because the energy from the rock bed has been significantly depleted and incoming cold fluid no longer can be heated to $\theta_f=1$ by the time it exits the storage tank.

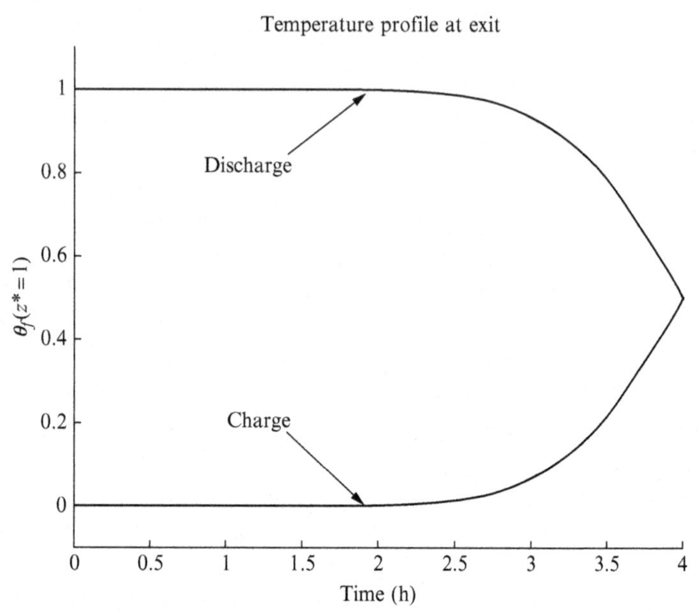

Fig. 4.9 Dimensionless temperature histories of the exit fluid at $z^*=1$ for charge and discharge processes.

The preceding numerical results agree with the expected scenario as described in Section 4.4. To validate this numerical method, analytical solutions were conducted using a Laplace transform method by the current authors [29], which were only possible for cases with a constant inlet fluid temperature and a simple initial temperature profile. Results compared in Fig. 4.10 are obtained under the same operational conditions—starting from a fully charged initial state and run for five iterations of cyclic discharge and charge processes. The fluid temperature distribution along the tank ($z^*=0$ for the bottom of the tank) from numerical results agrees with analytical results very well. This essentially proves the effectiveness and reliability of the numerical method developed in the present study.

As can be seen, the temperature distribution along z^* at the end of a charge is nonlinear. This distribution will be the initial condition for the next discharge cycle. Similarly, a discharge process will result in a nonlinear temperature distribution, which will be the initial condition for the next charge. It is evident that the analytical solutions developed by Schumann [17] could not handle this type of situation.

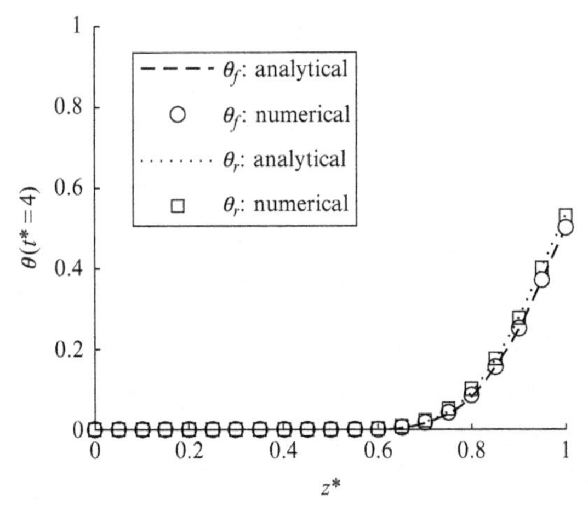

Fig. 4.10 Comparison of numerical and analytical results of the temperature distribution in the tank after time $t^*=4$ of a discharge (here θ_r is used to denote θ_s, as rocks are used as the storage material in the example).

Another special comparison was made to demonstrate the efficiency of the method of characteristics at solving the dimensionless form of the governing equations. Shown in Fig. 4.11 are the temperature profiles at $t^*=4$ obtained by using different numbers of nodes (20, 100, and 1000) for z^*. The high level of accuracy of the current numerical method, even with only 20 nodes, demonstrates the accuracy and stability of the method with minimal computing time.

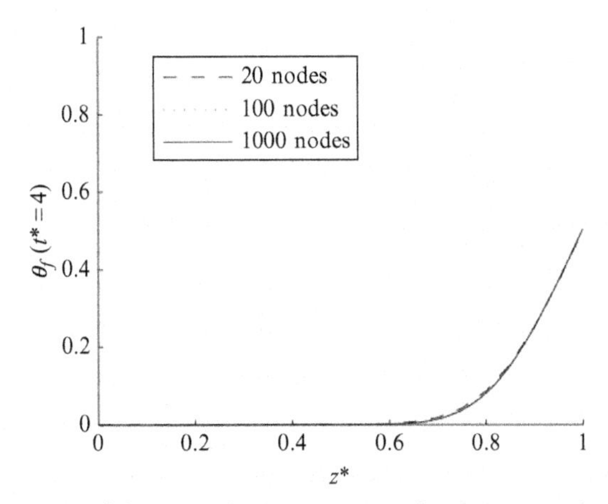

Fig. 4.11 Comparison of dimensionless temperature distributions in the tank after time $t^*=4$ of discharge for different numbers of discretized nodes.

4.3 ONE-DIMENSIONAL HEAT TRANSFER MODEL FOR ENCAPSULATED PCM PACKED BED

In this section, we considered the case of using encapsulated PCM as the storage material. Based on properties alone, the use of a PCM, combining both sensible and latent heat, allows a significantly higher energy storage density as compared to the use of sensible heat exclusively. Experimental studies of various tank filler materials confirm enhanced performance through use of a PCM, show resulting tank volume reduction by as much as a factor of 10, and suggest PCM fillers as fully viable alternatives for all thermal energy storage applications [30–33].

Efforts in thermal energy storage modeling go back as far as Schumann in 1929 [17], whose equations formed a basis for representing fluid flow through a porous packed bed thermal storage tank. Following models [18,23] expanded consideration, where most recently Van Lew et al. [25] applied the method of characteristics to produce a direct, fast, and accurate numerical solution to model thermocline interactions. A model by Regin and Solanki [34] considered a simple charge process of a tank with PCM filler for a parametric study of material properties. Following, a model by Wu et al. [35] applied an implicit finite difference method to solve the equations for the case with presence of PCM filler in the tank as a more general scenario, though results from the model featured numerous oddities and oscillations in temperature distribution profiles. To overcome the lower thermal conductivity of PCM material, Nithyanandam and Pitchumani [36,37] introduced heat transfer augmentation using thermosiphons or heat pipe. Different configurations were investigated by using computational fluid dynamics (CFD). Optimal orientation and design parameters were obtained. Archibold et al. [38] focused their attention on the fluid flow and heat transfer of the PCM within the spherical encapsulate. Recirculating vortexes were found in the upper region and therefore more intense melting occurs in this region. On the other hand, Vyshak and Jilani [39] used a modified enthalpy method to investigate the melting times for rectangular, cylindrical, and cylindrical shell storage configurations. The melting time was the least for cylindrical shell storage. They also investigated the effects of inlet temperature of the HTF. However, among these models, a comprehensive and accurate model for thermal energy storage with an encapsulated PCM filler has yet to be conceived.

The current work followed suit after the success of Van Lew et al. [25], with a much-needed expansion of analysis to an encapsulated PCM filler. An enthalpy-based version of the Schumann equations was used to allow tracking of interactions throughout the thermocline processes—a change especially necessary in the latent region where PCM filler temperature remained constant. The new set of equations was nondimensionalized for general application. With the resulting equations being of the hyperbolic type, the method of characteristics was applied for a numerical solution. The process gave fluid temperature and PCM filler enthalpy according to the discretized grid in time and space. With the equations following a similar form to those Van Lew obtained, we too expected the method to produce a direct solution that is both accurate and efficient.

The addition of enthalpy to consideration required an equation of state to close the gap in unknowns for solution. For proper application of this equation in the governing thermocline interactions, PCM filler phase states had to be tracked closely. More importantly, to maintain accuracy as these PCM filler phase states change throughout the space, a careful tracking of PCM filler phase state interfaces had to be implemented as well. This allowed proper application of the equations to all possible orientations and conditions of the PCM filler phase state interfaces in the numerical grid of characteristics. The method of characteristics made this possible, though the extent of generality and versatility hinged on the completeness of physical cases considered in its application.

4.3.1 Mathematical Model

To obtain the governing equations for fluid and PCM filler interactions in the thermocline, we first made some necessary assumptions. The work assumed a strictly vertical fluid flow through the tank, along with a uniform fluid distribution in the radial direction, as shown in Fig. 4.12. This reduced

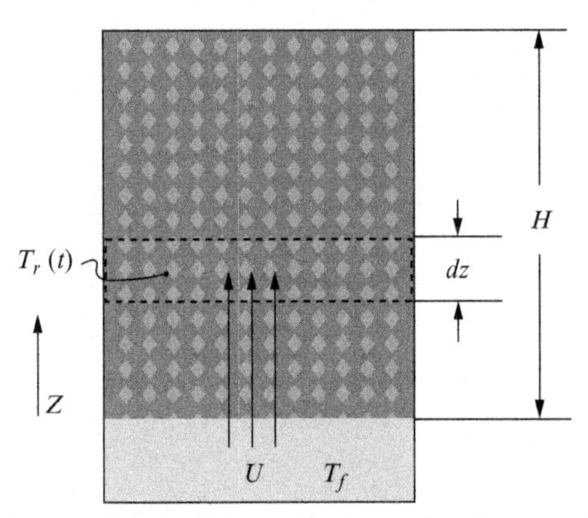

Fig. 4.12 General thermocline (pebbles or PCM capsules) for modeling.

consideration into a single spatial dimension z, which followed the direction of fluid flow. In this regard, z became a coordinate system that can be chosen, allowing identical application of the governing equations to both charge and discharge processes. Fluid thermophysical properties were assumed invariant with temperature, and thus constant. A general representation of the to-be-modeled thermocline tank can be viewed in Fig. 4.12.

With assumptions covered, we applied an energy balance to both fluid and filler for the chosen differential control volume. For the fluid, necessary terms included the enthalpy of the flow in and out of the volume, energy exchange between the fluid and PCM filler material within the volume, and the internal energy change of the fluid over the instance in time. The fluid governing equation became:

$$\rho_f C_f V_f \left(\frac{\partial T_f}{\partial t} + U \frac{\partial T_f}{\partial z} \right) = h S_s \left(T_s - T_f \right) \tag{4.62}$$

For the PCM filler, we needed only to consider the energy exchange between it and the fluid, and the change of internal energy of the PCM filler over the instance in time; see Tumilowicz et al. (2013). The PCM filler governing equation became

$$\rho_s V_s \frac{\partial \bar{h}_s}{\partial t} = -h S_s \left(T_s - T_f \right) \tag{4.63}$$

The preceding equations still retain a PCM filler temperature term, for which we applied an equation of state to relate to the enthalpy per PCM filler phase state.

$$T_s = \begin{cases} \dfrac{\bar{h}_s - \bar{h}_{s\_ref}}{C_{s\_s}} + T_{s\_ref} & \bar{h}_s < \bar{h}_{s\_melt} \\ T_{s\_melt} \quad \text{for} & \bar{h}_{s\_melt} < \bar{h}_s < \bar{h}_{s\_melt} + L \\ \dfrac{\bar{h}_s - (\bar{h}_{s\_melt} + L)}{C_{s\_\ell}} + T_{s\_melt} & \bar{h}_{s\_melt} + L < \bar{h}_s \end{cases} \tag{4.64}$$

For this enthalpy-based one-dimensional model, the lumped capacitance assumption is applied, considering the fact that the size of encapsulated fillers is small, which ensures a small Biot number. However, if the encapsulated filler material is large, which gives a large Biot number, the internal thermal resistance becomes significant, and a modification to the lumped capacitance has to be considered. This can be done by introducing an effective convective heat transfer coefficient in Eqs. (4.62) and (4.63). Detailed derivation of the formula of the effective heat transfer coefficient has been provided by Xu et al. [4].

For generality, dimensionless analysis was applied to the governing equations by introducing the following variables. Temperatures T_H and T_L are characteristic to the working system, representing the highest temperature inlet fluid used for heating during the charge process, and the lowest temperature fluid used for cooling during the discharge process, respectively.

$$\theta_f = \frac{T_f - T_L}{T_H - T_L} \tag{4.65}$$

$$\theta_s = \frac{T_s - T_L}{T_H - T_L} \tag{4.66}$$

$$t^* = \frac{t}{H/U} \tag{4.67}$$

$$z^* = \frac{z}{H} \tag{4.68}$$

$$\eta_s = \frac{\bar{h}_s - \bar{h}_{s\_ref}}{C_{s\_s}(T_{melt} - T_L)} \tag{4.69}$$

The governing equations in dimensionless form followed. Note that the subscripted "melt" and "r_ref" variables correspond to values plugged directly into their dimensionless variable (i.e., $\theta_s(T_s = T_{s\_melt}) \rightarrow \theta_{s\_melt}$).

$$\frac{\partial \theta_f}{\partial t^*} + \frac{\partial \theta_f}{\partial z^*} = \frac{1}{\tau_r}(\theta_s - \theta_f) \tag{4.70}$$

$$\frac{\partial \eta_s}{\partial t^*} = \frac{-H_{CR}}{\tau_r}\frac{1}{\theta_{s\_melt}}(\theta_s - \theta_f) \tag{4.71}$$

$$\theta_s = \begin{cases} \eta_s \theta_{s\_melt} + \theta_{s\_ref} & \eta_s < \eta_{s\_melt} \\ \theta_{s\_melt} \quad \text{for} & \eta_{s\_melt} < \eta_s < \eta_{s\_melt} + \dfrac{1}{Stf} \\ \left(\eta_s - \left(\eta_{s\_melt} + \dfrac{1}{Stf}\right)\right)\left(\dfrac{C_{s\_s}}{C_{s\_\ell}}\right)\theta_{s\_melt} + \theta_{s\_melt} & \eta_{s\_melt} + \dfrac{1}{Stf} < \eta_s \end{cases} \tag{4.72}$$

with dimensionless parameters defined as

$$\tau_r = \frac{U}{H}\frac{\rho_f C_f V_f}{h S_s} \tag{4.73}$$

$$H_{CR} = \frac{\rho_f C_f V_f}{\rho_s C_{s\_s} V_s} \tag{4.74}$$

$$Stf = \frac{C_{s\_s}\left(T_{s\_melt} - T_L\right)}{L} \tag{4.75}$$

4.3.2 Numerical Method and Procedures of Solution

To solve the equations presented, the method of characteristics was applied. Using an equal step size in both time and space $\Delta t^* = \Delta z^*$, we chose a numerical grid featuring both diagonal characteristics $t^* = z^*$ and vertical characteristics $z^* = constant$, as shown in Fig. 4.5. Steps in time progressed for $j = 1, 2, \ldots, N$, while steps in space progressed for $i = 1, 2, \ldots, M$.

Along the diagonal characteristic $t^* = z^*$, we recognized the substantial derivative in Eq. (4.70):

$$\frac{D\theta_f}{Dt^*} = \frac{1}{\tau_r}\left(\theta_s - \theta_f\right) \tag{4.76}$$

Here, we first saw the advantage of applying the method of characteristics to this hyperbolic system. By choosing the diagonal characteristic, the fluid equation reduced from a partial differential equation to an ordinary differential equation along this curve. Separating and integrating along the characteristic, we obtained:

$$\int d\theta_f = \int \frac{1}{\tau_r}\left(\theta_s - \theta_f\right) dt^* \tag{4.77}$$

We continued with a similar process for the PCM filler energy balance. Eq. (4.71) was solved along the characteristic $z^* = constant$. Again applying separation and integrating, we obtained

$$\int d\eta_s = \int \frac{-H_{CR}}{\tau_r} \frac{1}{\theta_{s\_melt}}\left(\theta_s - \theta_f\right) dt^* \tag{4.78}$$

Clearly, the result represented a system of ordinary differential equations along their corresponding characteristics. Referring back to the grid in Fig. 4.5, we see the two characteristics intersect as they progress in time and space. We exploited this in our manner of solution throughout the nodal grid. The hyperbolic nature of the original equations passes information from node to node in a wave-like fashion. As an example, we chose two neighboring spatial nodes at time $j = 1$, $\vartheta_{1,1}$ and $\vartheta_{2,1}$, which will serve as the starting points for information propagation through their corresponding characteristics. After the passing of one time step to $j = 2$, the meeting point of the two characteristics, $\vartheta_{2,2}$, will have received information from the two starting nodes. To represent this mathematically, we applied numerical integration to the equations. Along the diagonal characteristic, Eq. (4.77) became

$$\int_{\vartheta_{1,1}}^{\vartheta_{2,2}} d\theta_f = \int_{\vartheta_{1,1}}^{\vartheta_{2,2}} \frac{1}{\tau_r}(\theta_s - \theta_f)\, dt^* \tag{4.79}$$

Applying the trapezoidal rule for integration of the right-hand side, the solution to Eq. (4.79) became

$$\theta_{f_{2,2}} - \theta_{f_{1,1}} = \frac{\Delta t^*}{\tau_r}\left(\frac{\theta_{s_{2,2}} + \theta_{s_{1,1}}}{2} - \frac{\theta_{f_{2,2}} + \theta_{f_{1,1}}}{2}\right) \tag{4.80}$$

Repeating the process for Eq. (4.78), we obtained

$$\int_{\vartheta_{2,1}}^{\vartheta_{2,2}} d\eta_s = \int_{\vartheta_{2,1}}^{\vartheta_{2,2}} \frac{-H_{CR}}{\tau_r}\frac{1}{\theta_{s\_melt}}(\theta_s - \theta_f)\, dt^* \tag{4.81}$$

Once again implementing the trapezoidal rule for integration of the right-hand side, the integral solution became:

$$\eta_{s_{2,2}} - \eta_{s_{2,1}} = -\frac{\Delta t^* H_{CR}}{\tau_r \theta_{r\_melt}}\left(\frac{\theta_{s_{2,2}} + \theta_{s_{2,1}}}{2} - \frac{\theta_{f_{2,2}} + \theta_{f_{2,1}}}{2}\right) \tag{4.82}$$

Using the equation of state Eq. (4.72) to transform the unknown PCM filler temperature value at node $\vartheta_{2,2}$ to enthalpy based on the local PCM filler phase state left the system of two equations (4.80) and (4.82) to solve for the two unknowns, $\theta_{f_{2,2}}$ and $\eta_{r_{2,2}}$. With these values obtained, we stepped once in space to the new pair of neighboring nodes, $\vartheta_{2,1}$ and $\vartheta_{3,1}$, and used them identically to obtain values at node $\vartheta_{3,2}$. This was repeated until all values were found at $j=2$. We then fully repeated the spatial sweep at $j=2$ to obtain all new values at $j=3$.

Thus, with a boundary condition provided at the inlet $i=1$, along with an initial condition at time $j=1$ in the storage tank, solutions could be swept through space, stepped in time, and repeated, until the entire grid was fully calculated. Application of the trapezoidal rule for numerical integration implies accuracy of order $O(\Delta t^{*2})$ [28]. The preceding was a mere example calculation outlining the numerical process applied. With PCM filler phase state changes added, numerous technicalities had to be considered throughout the full application. While use of the state equation and solution of the system of equations followed simply in regions of continuous phase state, PCM filler phase state interfaces and their travel throughout the grid of characteristics created a plethora of more complex calculations. The necessary numerical considerations were most generally divided into calculations of type "boundary" and "spatial." Within these, we covered all cases of heating and cooling between different types of PCM filler phase states, PCM filler phase state interface presence/positioning, and the possibilities of overheating and undercooling. These more extreme cases were a necessary consideration with the application of the method of characteristics. In the chosen numerical grid, information could only propagate at the characteristic speed. With extreme heating or cooling conditions, an interface could be found traveling faster than the maximum allowed 45 degrees (the diagonal characteristic), at which point its position had to be suppressed to the maximum, and the additional energy transfer was accumulated in the resulting PCM filler enthalpy at the phase state interface.

At the boundary ($i=1$), only the vertical characteristic was used for calculation, and known fluid values allowed direct calculation at this point in space for all instances in time. Analysis of PCM filler enthalpy results revealed any resulting phase state changes in the filler. Overall, the model included a total of 11 considerations for grid calculations of this type: (1) continuous solid phase state, (2) continuous melting phase state, (3) continuous liquid phase state, (4) solidus phase state interface onset with heating, (5) solidus phase state interface onset with cooling, (6) liquidus phase state interface onset with heating, (7) liquidus phase state interface onset with cooling, (8) solidus phase state interface onset with overheating, (9) solidus phase state interface onset with undercooling, (10) liquidus phase state interface onset with overheating, and (11), liquidus phase state interface onset with undercooling.

To handle the discretized grids in the tank space, we worked with the system of equations resulting from the intersection of the two characteristics. Specific calculation depended on the phase state(s) present, and the phase state interface placement among the corresponding characteristic lines. To better detail the process, we considered a case where phase state interface travel during an instance in time leaves it within the same block in the discretized grid. With all values at the current time known, we first shifted the solution characteristics to solve for the phase state interface position at the new time, β_{New}, which represented a fractional positioning of the phase state interface in the grid $0<\beta<1$. The corresponding general solution space in the discretized grid can be viewed in Fig. 4.13.

The characteristics were discretized as before, with integration of the fluid equation from point 5 to the interface location β_{New}, and of the PCM filler enthalpy equation from point 6 to the interface location β_{New}. Application of linear interpolation (using the nearest known points) for intermediate terms 5 and 6, followed by elimination of the unknown fluid temperature at β_{New} from the system, resulted in a second-degree polynomial to solve for β_{New}. From the previously mentioned definition, $0<\beta<1$, selection of the proper root followed simply. With the position

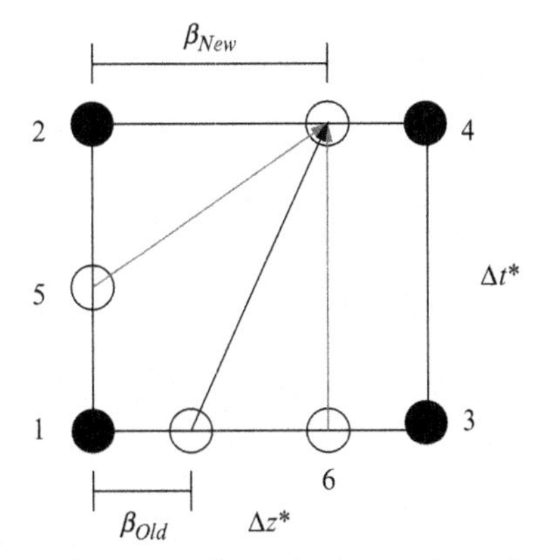

Fig. 4.13 Solution grid for first step of standard spatial interface travel.

at the new time found, the corresponding fluid value was obtained from the same characteristic equations.

With travel of the phase state interface now defined within the grid, we continued to define necessary values at the upper right point in the spatial sweep. Due to the presence of a phase state interface, the solution characteristic must have been divided accordingly. The general procedure is depicted in Fig. 4.14.

Linear interpolation was applied along the interface for values at its intersection with the diagonal characteristic. Finally, the system of equations defined by the two preceding characteristics was solved for the fluid and PCM filler temperatures at point 4.

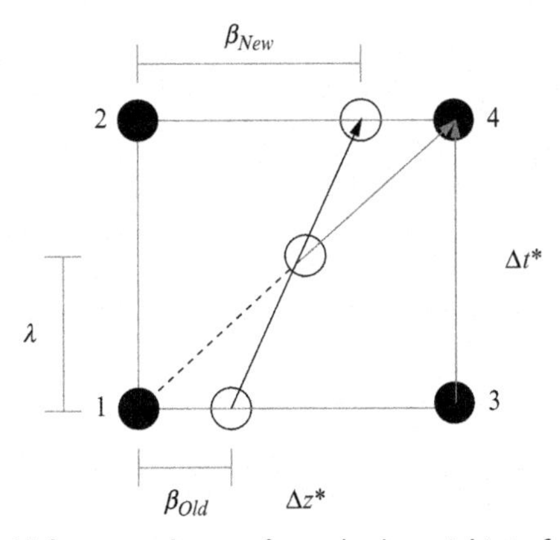

Fig. 4.14 Solution grid for second step of standard spatial interface travel.

Overall, the model included 19 considerations for grid calculations of spatial type: (1) continuous solid phase state, (2) continuous melting phase state, (3) continuous liquid phase state, (4) solidus phase state interface within grid with heating, (5) solidus phase state interface within grid with cooling, (6) liquidus phase state interface within grid with heating, (7) liquidus phase state interface within grid with cooling, (8) solidus phase state interface crossing vertical characteristic with heating, (9) solidus phase state interface crossing vertical characteristic with cooling, (10) liquidus phase state interface crossing vertical characteristic with heating, (11) liquidus phase state interface crossing vertical characteristic with cooling, (12) solidus phase state interface leaving spatial domain with heating, (13) solidus phase state interface leaving the spatial domain with cooling, (14) liquidus phase state interface leaving the spatial domain with heating, (15) liquidus phase state interface leaving the spatial domain with cooling, (16) solidus phase state interface crossing vertical characteristic with overheating, (17) solidus phase state interface crossing vertical characteristic with undercooling, (18) liquidus phase state interface crossing vertical characteristic with overheating, and (19) liquidus phase state interface crossing vertical characteristic with undercooling.

Extensive verifications were done to ensure the accuracy and robustness of the algorithm. The readers are referred to the details in the reference by Tumilowicz et al. [40]. Here, we will present a couple of examples to demonstrate the capabilities of the algorithm.

4.3.3 Examples of Results From Numerical Solution for Packed Bed of PCM Capsules

In this section, we briefly show some of the developed encapsulated PCM model's capabilities in application. We first consider a simple charge/discharge process for a thermocline utilizing this sort of material. We start with the charge process, with key numerical parameters set to

$$
\begin{aligned}
&H_{CR}=0.5785;\ \ \tau_r=0.1117;\ \ \theta_{f_{inlet}}=1;\ \ \theta_{f_o}=\theta_{s_o}=0;\ \ \theta_{s\_melt}=0.5\\
&\theta_{s\_ref}=0;\ \ \eta_{s\_melt}=1;\ \ \frac{C_{s\_s}}{C_{s\_\ell}}=1.1268;\ \ Stf=0.1143;\ \ \Delta t^{*}=\Delta z^{+}=0.001
\end{aligned}
\tag{4.83}
$$

The initially uniform temperature in the tank is heated by an HTF at a higher constant temperature. Heating occurred from the initial state to a final dimensionless time of 4.0, for which temperature profiles are displayed

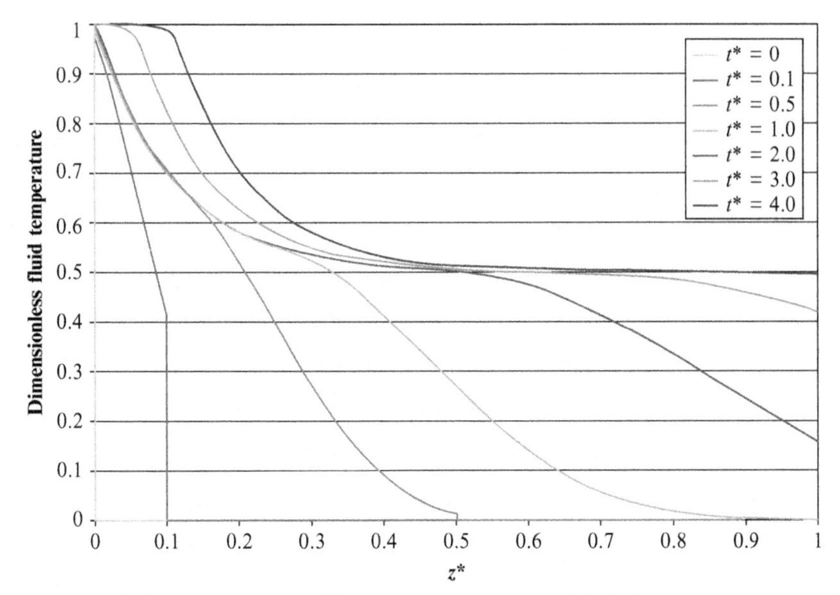

Fig. 4.15 Fluid temperature profiles at various instances of time for charge process of PCM filler.

at dimensionless times 0, 0.1, 0.5, 1.0, 2.0, 3.0, and 4.0. These can be viewed in Figs. 4.15 and 4.16, for the fluid and PCM filler, respectively.

The material behavior in the tank during the process can be observed. With the time allotted, the higher temperature HTF inflow raised the PCM filler temperature, melted it fully, and continued heating it in its liquid phase state. We see the PCM filler first started transitioning into its melting phase state, $\theta_{r\_melt} = 0.5$, soon after dimensionless time 0.1, and then finally collecting enough energy to continue to a liquid PCM filler just before dimensionless time 2.0.

Spatially, the positioning of the corresponding PCM filler phase states became quite clear. At the end of the charge, with PCM filler temperature at the outlet remaining just below the PCM filler melting temperature, heat propagation sent the solidus phase state interface near the end of the tank, leaving only a small amount of solid PCM filler near the outlet. The constant filler temperature of 0.5 throughout most of the space shows the large latent heat maintaining the PCM filler in this intermediate melting phase state. Finally, near the inlet, the temperature of the PCM filler finally rose above 0.5, signifying a liquid PCM filler, and placing the liquidus phase state interface near this inlet boundary. The interface positions as functions of time in Fig. 4.17 confirm this behavior.

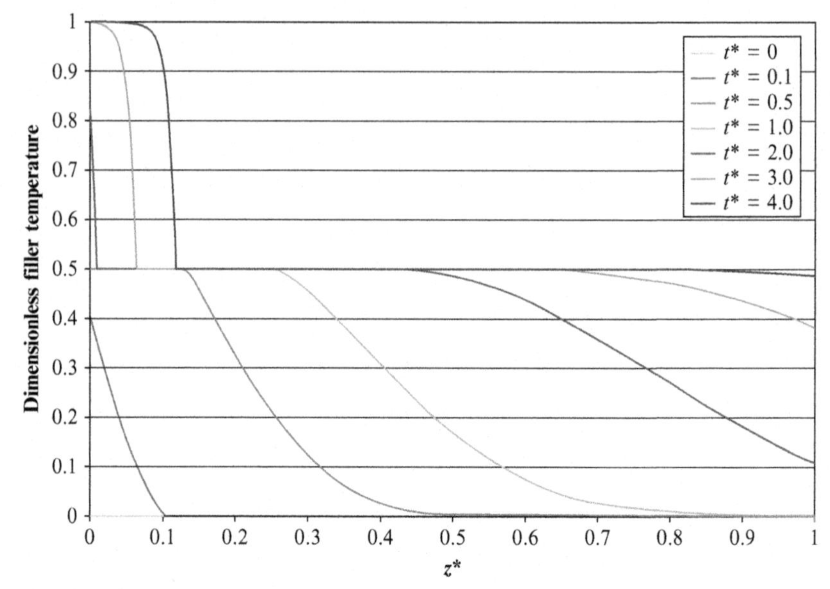

Fig. 4.16 Filler temperature profiles at various instances of time for charge process of PCM filler.

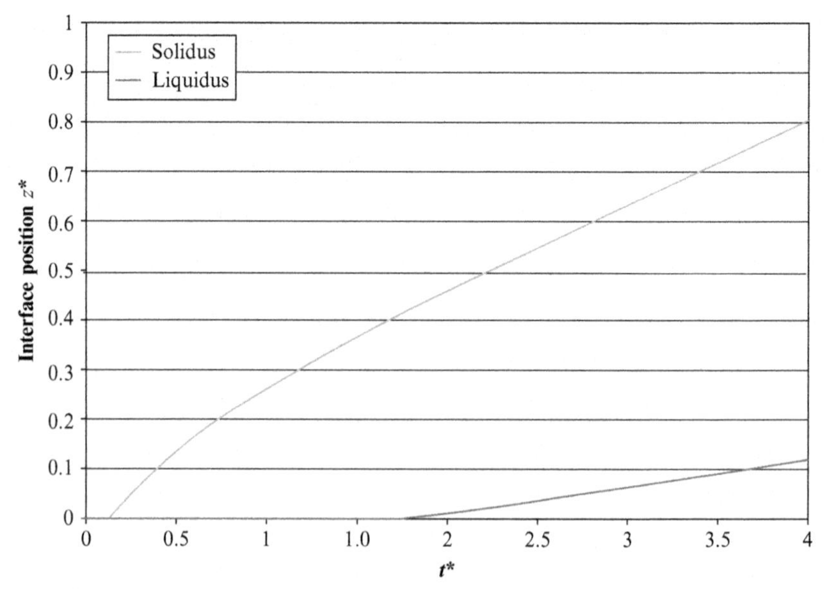

Fig. 4.17 Phase change interface positions as functions of time for standard charge process of PCM filler.

The plot confirms what is seen in the temperature distributions, in Fig. 4.16. Melting occurred relatively soon after heating began, corresponding to the start of the solidus phase state interface. With continued heating, this interface traveled through space, and is left near the end of the tank. Late in the charging cycle, enough heat had been input to produce a fully liquid PCM filler, corresponding to the beginning of a liquidus phase state interface. With only a little time left in the process, this second interface traveled only slightly into the space.

Postcharge process, we review how the tank achieves equilibrium to extract the effects of having a PCM filler. An equilibration calculation was applied to the final fluid and PCM filler temperature profiles before turning fluid flow around and discharging the stored energy. The final profiles of the tank, along with their corresponding equilibrium profile, can be viewed below in Fig. 4.18.

Reviewing the results, we see the effects of latent heat of equilibration results. With the end of charge profile featuring mostly melted PCM filler, the temperature difference between fluid and PCM filler was not enough to lift the PCM filler above its melting temperature. Thus, we saw a larger reduction in a relatively wide portion of the fluid temperature profile from equilibration, leading to a larger decrease of fluid output temperature for eventual energy production.

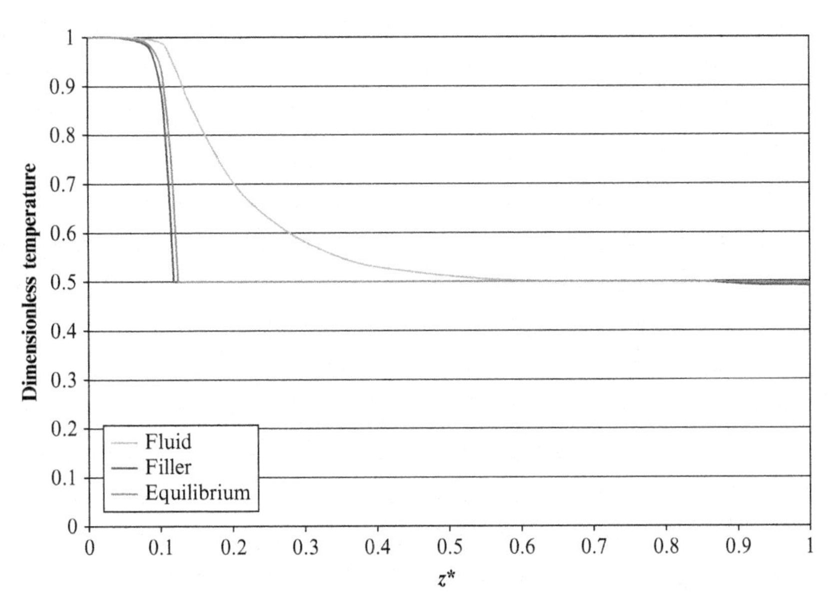

Fig. 4.18 Equilibration of final temperature profiles after charge process for PCM filler.

With equilibrium applied, the resulting temperature profile became the initial fluid and PCM filler condition for the discharge process. The tank was cooled at the inlet by a constant low fluid temperature, and key numerical parameters remained the same. Discharge was again run to a final dimensionless time of 4.0, for which the tank profiles are displayed at dimensionless time values 0, 0.1, 0.5, 1.0, 2.0, 3.0, and 4.0. Results are plotted in Figs. 4.19 and 4.20 for the fluid and PCM filler, respectively.

For this discharge process, the profiles behaved in a reverse order from the charge. Cooling brought the PCM filler temperature down, reduced the amount of liquid PCM filler, and increased the amount of solid PCM filler. At dimensionless time 1.0, cooling sent the liquidus phase state interface out of the storage tank, thus solidifying all liquid PCM filler. Heat extraction continued until the solidus phase state interface traveled past the position 0.8, leaving mostly solid PCM filler, with a comparatively small amount of melting PCM filler. The fluid temperature profiles followed as expected, with discontinuities between inlet affected fluid and initial state seen before dimensionless time 1.0, as before.

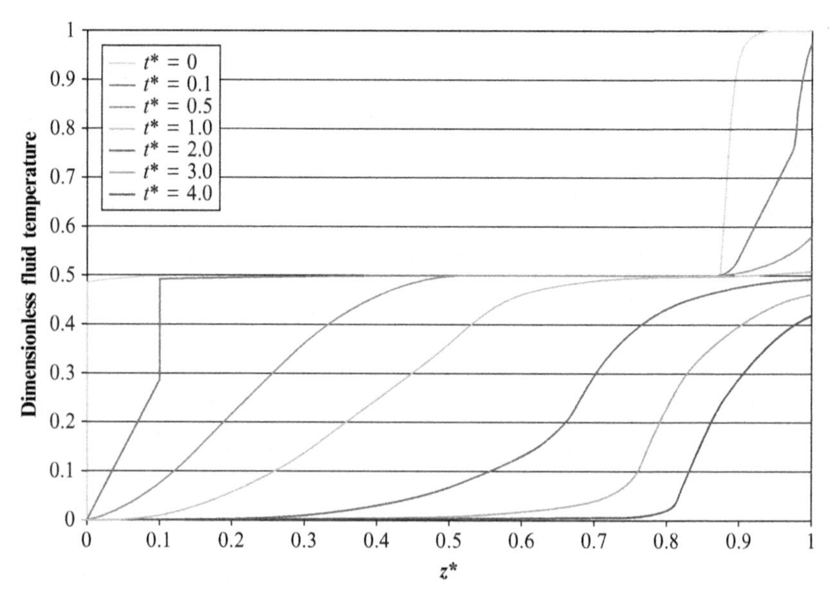

Fig. 4.19 Fluid temperature profiles at various instances of time for discharge process for PCM filler.

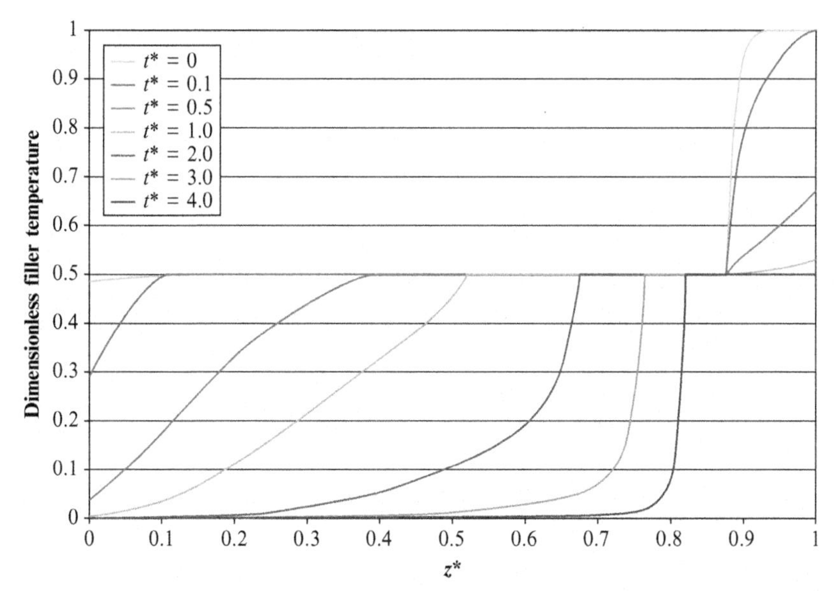

Fig. 4.20 Filler temperature profiles at various instances of time for discharge process for PCM filler.

Variable Heat Transfer Fluid Temperature at the Inlet

We showcase the versatility of the model by applying a variable boundary condition. With it, we aimed to apply heat to the PCM filler, forcing phase change(s), at which time the boundary condition was reversed, removing heat to produce an additional phase state interface reverting to the earlier phase state(s). Seeing satisfactory modeling in a complex heat transfer case such as this will confirm the general consideration implemented. The model was run as a charge process until a final dimensionless temperature equal to 5.0. Key properties were as follows:

$$\begin{aligned} &H_{CR} = 0.5785;\ \ \tau_r = 0.1117;\ \ \theta_{f_{inlet}} = 1; \\ &\theta_{f_o} = \theta_{s_o} = 0;\ \ \theta_{s\_melt} = 0.1;\ \ \theta_{s\_ref} = 0; \\ &\eta_{s\_melt} = 1;\ \ \frac{C_{s\_s}}{C_{s\_\ell}} = 1.1268;\ \ Stf = 0.1143; \\ &\Delta t^* = \Delta z^+ = 0.001 \end{aligned} \tag{4.84}$$

with inlet fluid temperature as follows.

$$\theta_{s}\_inlet = \begin{cases} 1.0 & & 0 \le t^* < 1.5 \\ 0.6 & \text{for} & 1.5 \le t^* < 2.5 \\ 0.3 & & 2.5 \le t^* < 4.0 \\ 0 & & 4.0 \le t^* < 5.0 \end{cases} \tag{4.85}$$

This boundary condition represented a discontinuous series of constant fluid temperature values applied over specific amounts of time. Heating was applied quickly to produce two phase state interfaces, and then a gradually stepped cooling was used to revert to a previous phase state. Below, we trace the temperature profile evolutions throughout the various changes in boundary condition.

We first look at the temperatures within the tank at dimensionless time 1.499, which is the last instance of inlet fluid condition equal to 1.0. This can be viewed in Fig. 4.21. The profiles produced represent a simple constant inlet PCM filler temperature charge. Heating produced two phase state interfaces, leaving three separate phase states of PCM filler in the tank. Though paraffin features a large latent heat, the large temperature gradient left above the melting temperature due to the high inlet fluid temperature heated through the melting PCM phase state relatively quickly. We now continue to the distributions at dimensionless time 2.499, Fig. 4.22, which is the last instance in time with the inlet temperature of 0.6 as the boundary condition.

At this instant in time, we see an interesting response in the profiles. Due to the wide spread of PCM filler temperature after the previous heating, the new inlet condition worked to both cool and heat, depending on the inlet

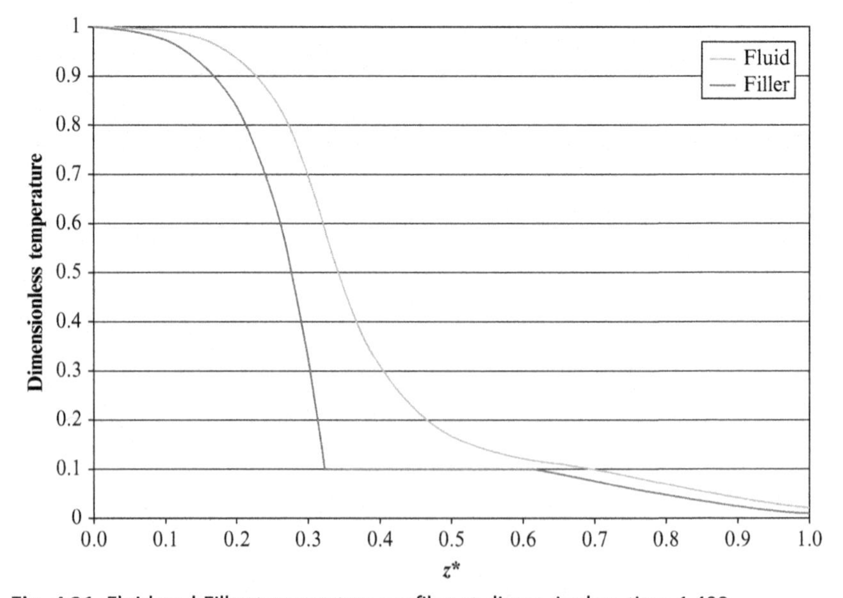

Fig. 4.21 Fluid and Filler temperature profiles at dimensionless time 1.499.

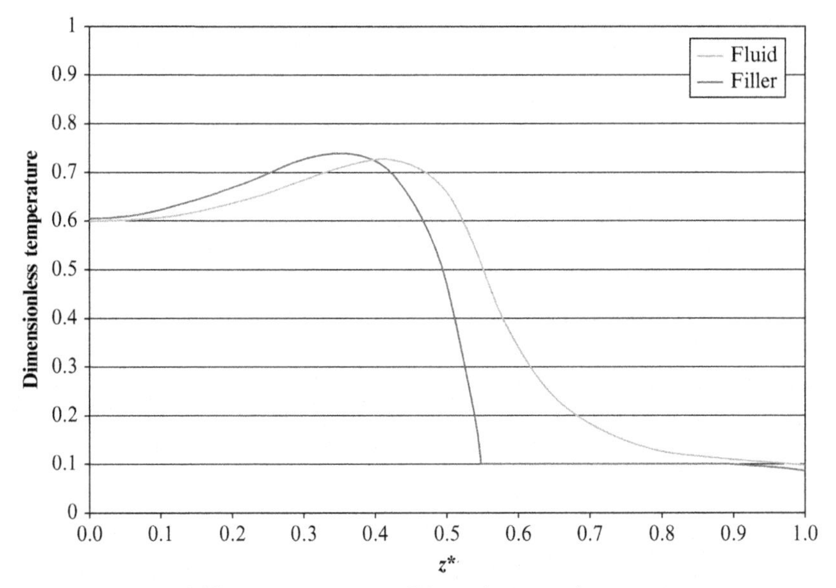

Fig. 4.22 Fluid and filler temperature profiles at dimensionless time 2.499.

fluid's relation to the temperatures within the tank at a certain position. The effect was seen throughout the entire tank, as this new condition, traveling at the characteristic speed, was able to traverse the entire space in the $\Delta t^* = 1.0$ it was applied. When interacting with the high temperatures towards the tank inlet, we saw a cooling that brings both PCM filler and fluid profiles gradually down to the inlet temperature condition. Further in space, where tank temperature values remain lower, the new condition continued to heat the tank profiles. The two heat transfer types met in a continuous parabolic fashion, representing the eventual change of liquid PCM filler being heated to liquid PCM filler experiencing cooling. Heat propagation through the tank had also facilitated the propagation of the two previously mentioned interfaces, as expected. The initial solidus phase state interface was found very close to the outlet, leaving little solid PCM filler in the tank, while the creation of more liquid PCM filler placed the liquidus phase state interface past the midway point in the tank.

Past this time, another lower fluid boundary temperature was applied up until the dimensionless time 3.999. The profiles at this time can be viewed in Fig. 4.23. As expected, this new inlet fluid temperature dropped temperature profiles even lower. All solid PCM filler has now melted, meaning the solidus phase state interface has left the space at some time during the application of this inlet condition. Conditions had also produced even more

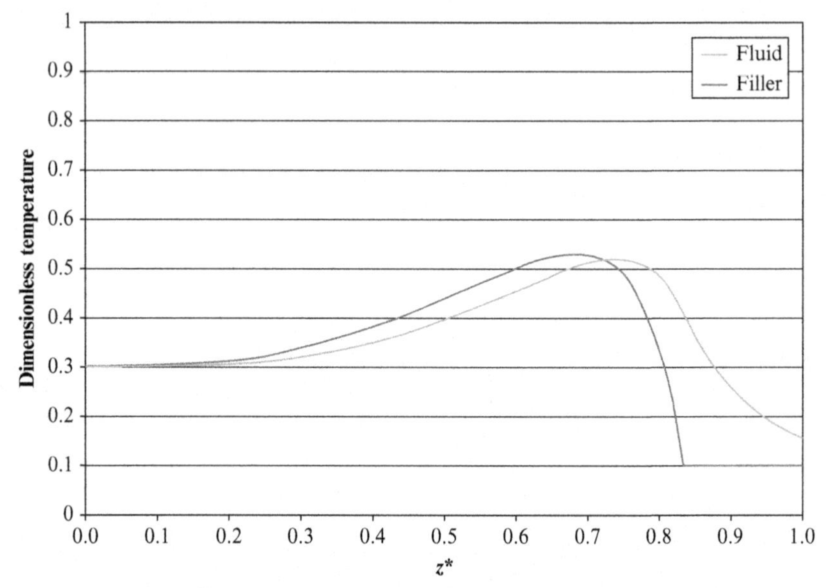

Fig. 4.23 Fluid and filler temperature profiles at dimensionless time 3.999.

liquid PCM filler, and correspondingly shifted the liquidus phase state interface further towards the outlet. Finally, the last drop in boundary temperature was applied from dimensionless time 4.0 through the end of the run. The resulting end temperature profile can be viewed in Fig. 4.24.

With this inlet fluid temperature, the system saw the lowest cooling temperature in the run. The PCM filler temperature was even further decreased, eventually returning the PCM filler at the boundary back to a melting PCM filler, and creating a second liquidus phase state interface. The original melting PCM filler was almost all fully liquefied, meaning we expected the first liquidus phase state interface to be very close to the tank outlet. This left three unique phase states in the tank from input to output—melting, liquid, and melting. To complete our understanding of the process, we compiled the travel of interfaces throughout. This can be viewed in Fig. 4.25.

The interface travel confirms what was outlined in the preceding profiles. The first phase state interface, of solidus type, was created early in the heating. Slightly past the halfway mark of the process, this phase state interface traveled far enough to leave the space entirely, taking all solid PCM filler with it. After heating through the latent region of melting PCM filler, the first liquidus phase state interface was created, bringing with it liquid PCM filler to the tank. This phase state interface traveled nearly the

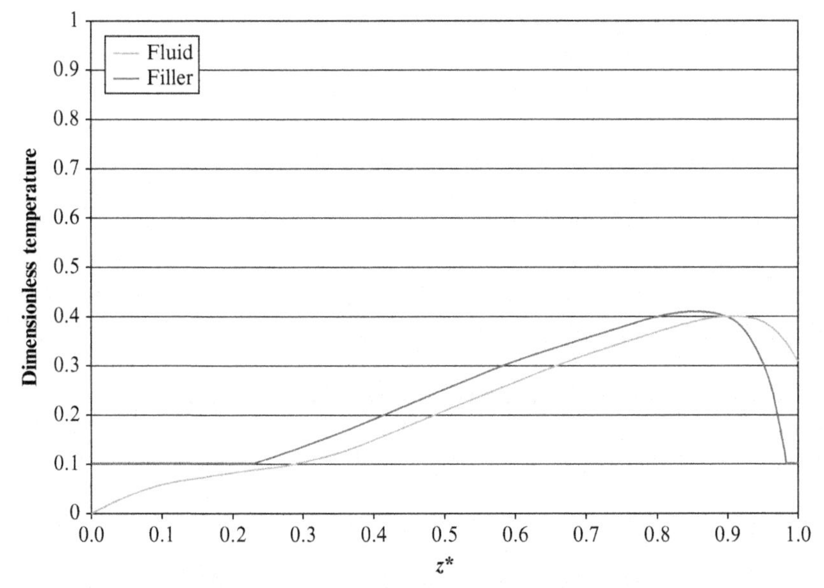

Fig. 4.24 Fluid and filler temperature profiles at dimensionless time 5.0.

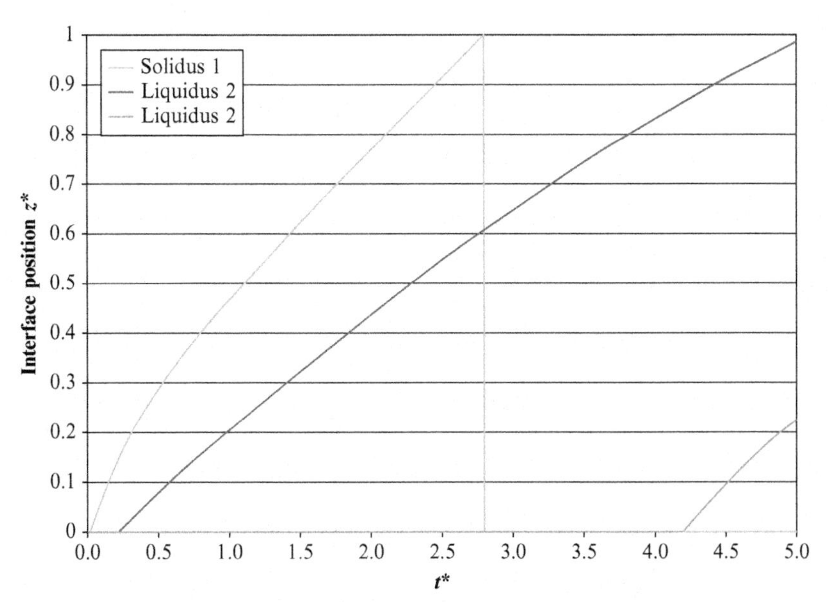

Fig. 4.25 Phase change interface positions as functions of time for variable boundary charge process.

entire space, and turned nearly all the remaining melting PCM filler into liquid PCM filler. Finally, after adequate cooling, a second liquidus phase state interface was created as the liquid PCM filler began returning to a melting PCM filler. This interface saw little travel, as this phase change occurred near the end of the run. Thus, throughout the entire case, heating and cooling created three unique phase state interfaces, all of which were tracked by the model. Interface travel was clearly nonlinear, though speed decreased to a near constant rate as the temperature difference between fluid and PCM filler driving their travel decreased with time.

4.4 VALIDITY OF ONE-DIMENSIONAL MODELS OF SENSIBLE AND LATENT HEAT THERMAL STORAGE

4.4.1 Sensible Thermal Storage

In Section 4.2.3 we presented the derivations of four different geometries. Table 4.1 lists the formulas of the effective heat transfer coefficient h_{eff}, on the basis of the intrinsic heat transfer coefficients h, for the four fluid-solid structural combinations shown in Fig. 4.3. All four solid-fluid configurations are generally viewed as systems with an HTF flowing through a porous medium, each should have porosity, or equivalent porosity for cases in Fig. 4.3B–D, defined as the ratio of the volume of fluid over the total volume of the storage tank. The governing equations of the energy balance for the fluid and solid are given in Section 4.3.4 and Eqs. (4.19) and (4.20). The hot HTF flows downward through the packed bed during heat charging, and the cold HTF flows upward during heat discharging. The fluid inlet (at the top during charging and at the bottom during discharging) always has $z=0$. The flow velocities are assumed to have a uniform radial distribution, which Yang and Garimella [41] have proven to be reasonable for packed beds.

In cases where the solid material is not in the form of spheres packed, as shown in Fig. 4.3A, but in a form such as those shown in Fig. 4.3B–D, equivalent porosity must be used. With equivalent porosity, the same governing equations can be used; however, an effective heat transfer coefficient h_{eff}, as given in Table 4.1, must be used to replace the original (or intrinsic) heat transfer coefficient h in Eqs. (4.19) and (4.20).

The parameter S_s in the governing equations, Eqs. (4.19) and (4.20), is the total surface area of solid thermal storage material in contact with HTF per unit length of the thermal storage tank (see Li et al. [42] for details on obtaining this parameter). As a consequence, the heat transfer and energy balance in the solid-fluid structural combinations in Fig. 4.3B–D can all be analyzed by only considering one typical volume, which includes a typical solid and fluid region as indicated on the right side of each of the configurations in Fig. 4.3. For the case of plates, the length in the direction normal to the paper in Fig. 4.3B is chosen to be a unit length of $L = 1.0\,m$. As a result, the flow channel will have a ratio of $L/D_f > 20$, where D_f is the width of the flow channel. Table 4.4 gives the obtained S_r for the other three cases in Fig. 4.3.

Introducing the following dimensionless variables,

$$\theta_f = (T_f - T_L)/(T_H - T_L) \tag{4.86.a}$$

$$\theta_r = (T_{LM} - T_L)/(T_H - T_L) \tag{4.86.b}$$

$$z^* = z/H \tag{4.86.c}$$

$$t^* = t/(H/U) \tag{4.86.d}$$

the governing equations become

$$\frac{\partial \theta_f}{\partial t^*} + \frac{\partial \theta_f}{\partial z^*} = \frac{1}{\tau_r}(\theta_r - \theta_f) \tag{4.87}$$

$$\frac{\partial \theta_r}{\partial t^*} = -\frac{H_{CR}}{\tau_r}(\theta_r - \theta_f) \tag{4.88}$$

where

$$\tau_r = \frac{U \rho_f C_f \varepsilon \pi R^2}{H \ h_{eff} S_{filler}} \tag{4.89}$$

and

$$H_{CR} = \frac{\rho_f C_f \varepsilon}{\rho_r C_r (1-\varepsilon)} \tag{4.90}$$

Under the assumption of no heat loss from the thermal storage tank, it is reasonable that the equilibrium temperature between the HTF and the solid

Table 4.4 The total heat transfer surface area of a solid per unit height of a typical volume as shown in Fig. 4.3B–D

	Total heat transfer surface area (m^2)	Height (m)	S_{filler} (m)
Plate	$2(1 \times H)$	H	2
Cylinder	$2\pi r H$	H	$2\pi r$
Tube	$2\pi a H$	H	$2\pi a$

filler material at the end of one charge or discharge will necessarily be the initial condition of the next discharge or charge process in the thermal storage cycle. This connects the discharge and charge processes so that results of a number of periodic charges and discharges can be obtained.

For the initial condition of fluid and filler material in any charging or discharging process, $t^* = 0$; $\theta_r = \theta_f$, which is the equilibrium state after settling down from the last process.

For the inlet condition, $z^* = 0$ and $\theta_f = 1$ for a charging process; otherwise $\theta_f = 0$ for a discharging process.

The θ_r at the inlet boundary can be directly calculated using Eq. (4.88) from the known inlet fluid temperature.

A pair consisting of an HTF and a thermal storage material has been chosen for the one-dimensional model as well as the CFD study for the purpose of comparison and verification. The high temperature of fluid in the charge process is 390°C (663.15 K) and the low temperature of fluid flowing into the tank during a discharge is 310°C (583.15 K). The HTF is HITEC molten salt [43], having properties of kinetic viscosity $\nu_f = 1.17 \times 10^{-6} \text{m}^2/\text{s}$, heat capacity $Cp_f = 1549.12 \text{J/kgK}$, density $\rho_f = 1794.07 \text{kg/m}^3$, and thermal conductivity k_f=0.57 W/m K. The thermal storage material is another molten salt with properties of $\rho_r = 1680 \text{kg/m}^3$, $C_r = 1560 \text{J/(kgK)}$, and $k_r = 0.61 \text{W/(mK)}$. The required time period for energy charge and discharge is 4 h each.

The three cases of solid-fluid structural combinations, as shown in Fig. 4.3B–D, were calculated using the method of characteristics based on the 1D model. For the convenience of defining the dimensions of the solid and liquid area, as well as the computational domain for CFD analysis, an equivalent control volume was defined. For the bundle of solid rods or the fluid tubes in Fig. 4.3C and D, we may consider that each rod or tube has a control area enclosed by a hexagon, as shown in Fig. 4.26. The outer boundary surface is thermally insulated. In the CFD analysis, the hexagon is equivalent to a circle in the same area. A comprehensive CFD analysis was performed by Li et al. [44] in which they found that the energy storage is the same in both hexagon and circle geometries.

As shown in Fig. 4.26, the equivalent diameter for the control area should satisfy the relationship

$$A_{hexagon} = A_{eq} = \pi R_{eq}^2 \tag{4.91}$$

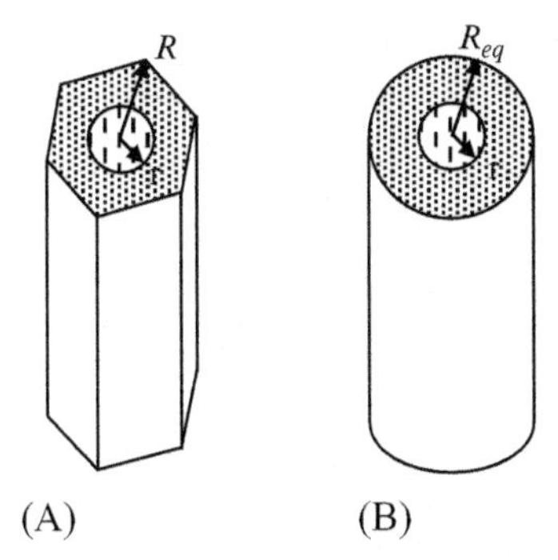

Fig. 4.26 (A) Hexagonal control volume; (B) equivalent circular control volume.

The equivalent diameters of the control area for the solid rod and the fluid tube are given in Table 4.5, which essentially defines the dimensions of flow channels and solid thermal storage materials. The height of the tank is 10 m. The related intrinsic heat transfer coefficient of laminar flow and the corrected heat transfer coefficient in the 1D model simulations are listed in Table 4.6.

The laminar flow heat transfer Nusselt numbers listed in Table 4.6 are based on heat transfer cases with constant wall heat flux [2] for fully developed internal flows. The full thermal and hydraulic development can be easily satisfied due to the very long channels. The assumption of a constant wall heat flux heat transfer is because the temperature difference between the solid materials and the fluid along the height of a storage tank is similar to that of a countercurrent flow heat exchanger—hot fluid charges into the tank from the top, where there is a high temperature, and during a discharge cold fluid flows into the tank from the bottom of the tank, where there is a low temperature. A detailed description of this variation of temperatures of fluid and solid during energy charge and discharge is given by Li et al. [45]. The Biot numbers of all cases of heat conduction in solid materials in this analysis are above 1.0.

The charging and discharging cyclic operations start with a charge to a cold tank. After several cyclic runs, the temperature distribution in a tank after a discharge becomes independent of the initial temperature distribution. This state is called a cyclic steady state, and it is the real situation in a solar power plant practical operation.

Table 4.5 Dimensions of the fluid channels and solid thermal storage structure

	Equivalent porosity ε	**Equivalent outer diameter** D_{eq} **(mm)**	**Diameter** d **(mm)**	S_r	**Fluid channel**[a] $2x_1$ **(mm)**	**Solid plate**[a] $2(x_2 - x_1)$ (mm)
Plate case	0.33	No	No	2.0	4.56	9.25
Cylinder case	0.33	43.5	35.6	0.1118	No	No
Tube case	0.33	43.5	25.0	0.0785	No	No

[a]See definition of x_1, x_2 in Fig. 4.3B.

Table 4.6 Parameters in the simulation of 1D transient model

Hydraulic diameter for Re and Nu	Nu number (constant heat flux)	Intrinsic heat transfer coefficient h (W/m^{2}K)	Effective heat transfer coefficient h_{eff}(W/m^{2}K)	τ_r	H_{CR}
Plate case ($D_h = 4x_1$)	8. 24	305	221	0.0039	0.5223
Cylinder case ($D_h = D_{eq} - d$)	5.18 (inner wall) (outer wall no heat flux)	352	99	0.0087	0.5223
Tube case ($D_h = d$)	4.36	94	64	0.0366	0.5223

The average fluid temperatures at height locations along the tank after a charging are shown in Fig. 4.27 for the number of cycles of charging/discharging as indicated by the legend. The velocity of HTF flowing into the tank for the following cases is $V_{inlet} = 1.36 \times 10^{-3}$m/s. The results become the same with no more changes in charge/discharge cycles after a certain number of cycles for all the cases. For the case of channels formed by flat plates, 11 cycles are required to reach the cyclic steady state; this number for the cylinder case and tube case is 10 and 7, respectively. Because the 1D transient model is rather convenient, it can simulate many cycles of charging/discharging in a relatively short computational time. This is the advantage of the 1D transient model over a CFD analysis, which would take a significant amount of computational time for the analysis of cyclic charges and discharges.

A comprehensive CFD study was employed to analyze the energy storage process and validate the 1D simplified transient model. The commercial software ANSYS Fluent@ 6.3 was chosen for this analysis, and GAMBIT 2.4 was used to generate the computational domain and grid system.

Computational Specifications

The flow and heat transfer in the thermal storage tank are incompressible and transient with constant properties. The fluid flow Reynolds numbers are usually in the laminar region. In the current three configurations of solid-fluid in storage tanks, the velocities of the HTF all equal $V_{inlet} = 1.36 \times 10^{-3}$m/s, according to the desired mass flow rate. Correspondingly, the Reynolds numbers are: $\mathrm{Re}_{plate} = 10.6$, $\mathrm{Re}_{tube} = 28.9$, and $\mathrm{Re}_{cylinder} = 9.2$. The hydraulic

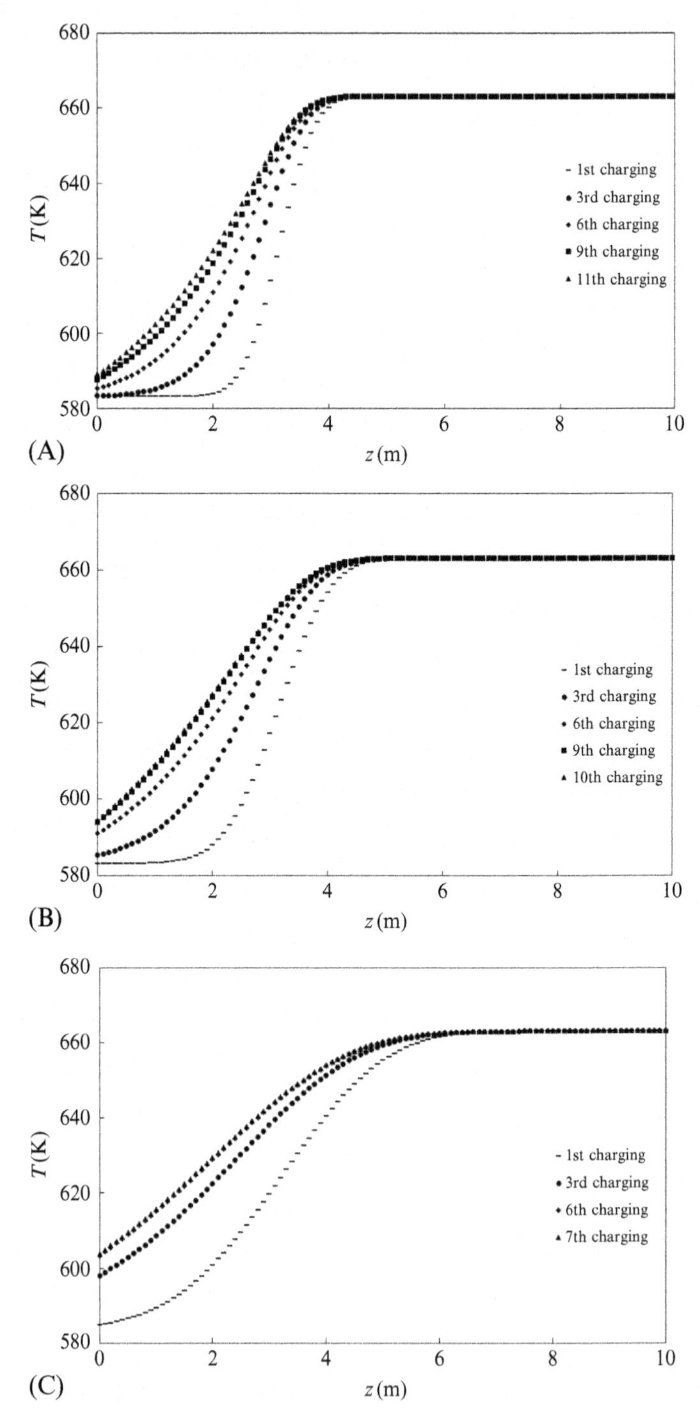

Fig. 4.27 The temperature of HTF at locations from bottom to top in a tank after heat charging processes. (A) Flat plate case, (B) cylinder case(C) tube case.

diameters defined in Table 4.6 for the flow channels were used to calculate these Reynolds numbers.

The continuity, momentum, and energy equations in differential form for laminar and incompressible flow are:

$$\nabla \cdot \vec{V} = 0 \tag{4.92}$$

$$\frac{\partial \vec{V}}{\partial t} + \left(\vec{V} \cdot \nabla\right)\vec{V} = -\frac{1}{\rho}\nabla p + \nu\nabla^2\vec{V} \tag{4.93}$$

$$\frac{\partial T}{\partial t} + \left(\vec{V} \cdot \nabla\right)T = \kappa\nabla^2 T \tag{4.94}$$

Because of the thermocline effect (hot fluid on top of cold fluid) in thermal storage tanks, typically there is no natural convection to consider. The computational domains and boundary conditions of the three cases are described in the following:

(1) For the solid plates and the 2D channels, as shown in Fig. 4.3B, half of the flow channel and half of the plate were included in the computational domain. At the symmetric line of the flow channel, $u_x = 0$, $\frac{\partial u_z}{\partial x} = 0$, and $\frac{\partial T}{\partial x} = 0$. At the symmetric line of the solid plate, $u_z = 0$, $u_x = 0$, and $\frac{\partial T}{\partial x} = 0$. At the solid walls of $z = 0$ and $z = H$, $u_z = 0$, $u_x = 0$, and $\frac{\partial T}{\partial z} = 0$. Full development conditions at the outflow boundary are used, as the flow channels are sufficiently long. This gives $u_x = 0$, $\frac{\partial u_z}{\partial z} = 0$, $\frac{\partial T}{\partial z} = 0$.

(2) For the solid rod with fluid flowing along the length, as shown in Fig. 4.3C, the rod and the equivalent circular area around the rod are included in the computational domain. The boundary conditions include: at the centerline of the rod, $u_z = 0$, $u_r = 0$, and $\frac{\partial T}{\partial r} = 0$. At the outer boundary, in the r direction, $u_r = 0$, $\frac{\partial u_z}{\partial r} = 0$, and $\frac{\partial T}{\partial r} = 0$. At the solid walls of $z = 0$ and $z = H$, $u_z = 0$, $u_r = 0$, and $\frac{\partial T}{\partial z} = 0$. At the outflow boundary, fully developed boundary conditions are: $u_r = 0$, $\frac{\partial u_z}{\partial z} = 0$, and $\frac{\partial T}{\partial z} = 0$.

(3) For the tubes surrounded by solid areas, as shown in Fig. 4.3D, a tube and the equivalent circular solid area around the tube are included in the computational domain. At the centerline of the tube, $u_r = 0$, $\frac{u_z}{\partial r} = 0$, and $\frac{\partial T}{\partial r} = 0$. At the outer boundary, in the r direction, $u_z = 0$, $u_r = 0$, and $\frac{\partial T}{\partial r} = 0$. At the solid walls of $z = 0$ and $z = H$, $u_z = 0$, $u_r = 0$, and $\frac{\partial T}{\partial z} = 0$. At the outflow boundary, fully developed boundary conditions are: $u_r = 0$, $\frac{\partial u_z}{\partial z} = 0$, and $\frac{\partial T}{\partial z} = 0$. The fluid and solid interfaces inside the computational domain have conjugated heat transfer, which can be typically treated in the software package ANSYS Fluent@.

Grid and Time Step Independent Study

A grid-independent study was conducted to choose a grid number that ensured high accuracy of computational results at a reasonable computational load. The flow and heat transfer fields were computed in 2D with cell numbers of 10,000, 20,000, 40,000, 60,000, and 80,000. For this study, we also assume the initial condition that the thermal storage tank is fully charged to a high temperature of 663.15 K, and the cold HTF with constant temperature 583.15 K flows into the tank with a constant velocity of $V_{inlet} = 1.36 \times 10^{-3}$m/s to extract the heat for 4 h. The temperature of the HTF at 1.0 m downstream of the inlet was examined for the transient process. It was found that the relative difference of the temperature at this location and any time instance during 4 h has a variation of no more than 10% when increasing the cell number from 10,000 to 20,000. The variation went down to 2.5% from 20,000 to 40,000, and 0.86% from 40,000 to 60,000, and 0.33% from 60,000 to 80,000. Consequently, 60,000 cells were adopted for the computational domains in all the formal computations of the study.

The time step for the transient flow field computation was set as 2 s, based on the time-step independence analysis. Time steps of 0.5, 1, 2, 4, and 5 s were tested in the computations. It was found that the relative difference of the temperature of the HTF at the location of 1 m from the inlet had a variation of 0.16% when increasing the time step from 0.5 to 1 s. The variation was 0.83% from 1 to 2 s, 1.95% from 2 to 4 s, and 3.67% from 4 to 5 s. Consequently, a small time step of 2 s was chosen for the computation of the transient process. At each time step of computation, convergence was checked to meet a convergence criterion.

Comparison of CFD Results and the Results From 1D Modeling

The temperature of the HTF at the exit of a discharge is an important indication of energy storage efficiency. Therefore, the HTF temperature at the exit during the 4 h discharging period has been monitored for comparison between the CFD results and the 1D modeling results. The results are for the cases after a sufficient number of cyclic chargings and dischargings, thus achieving cyclic steady state. In any 4 h charge process, the entering hot HTF has a constant temperature of 663.15 K, and in any discharge the exiting cold HTF has a constant temperature of 583.15 K. The inlet flow velocity of the fluid is always set as $V_{inlet} = 1.36 \times 10^{-3}$m/s.

As shown in Fig. 4.28, the fluid temperatures in the 4 h discharge process between the results of CFD and the 1D transient model agree very well. The maximum difference between them is 0.06% for the plate case, 0.05% for the cylinder case, and 0.01% for the tube case. This comparison clearly shows that the 1D transient model, which predicted fluid temperature at the exit of the discharge process during the time period, is sufficiently accurate. Engineers can use this method for the analysis of thermal energy storage systems without using comprehensive CFD computations.

To further compare the results from the 1D transient modeling with the results from CFD computation, Fig. 4.29 shows the distribution of temperature of the HTF at locations in the storage tank along the flow direction after 4 h of heat discharge. The heat discharge results are from a typical process after many cyclic charges and discharges so that a cyclic steady state has been achieved. It is understandable that after discharge, the temperature of fluid at the lower part of the tank is low. The temperature distribution from the 1D transient model and from the CFD computation agree very well, which further verifies the accuracy of the 1D transient modeling.

In conclusion, it is evident that the 1D transient model with the corrected heat transfer coefficient is robust and highly accurate for analysis of the heat transfer and energy storage behavior for the three studied typical fluid-solid structural combinations.

Because the 1D transient model significantly simplifies analysis, and especially the computational time, it is expected to be one of the most convenient and accurate tools available for industrial engineers to analyze the behavior of thermal energy storage systems of various fluid-solid configurations.

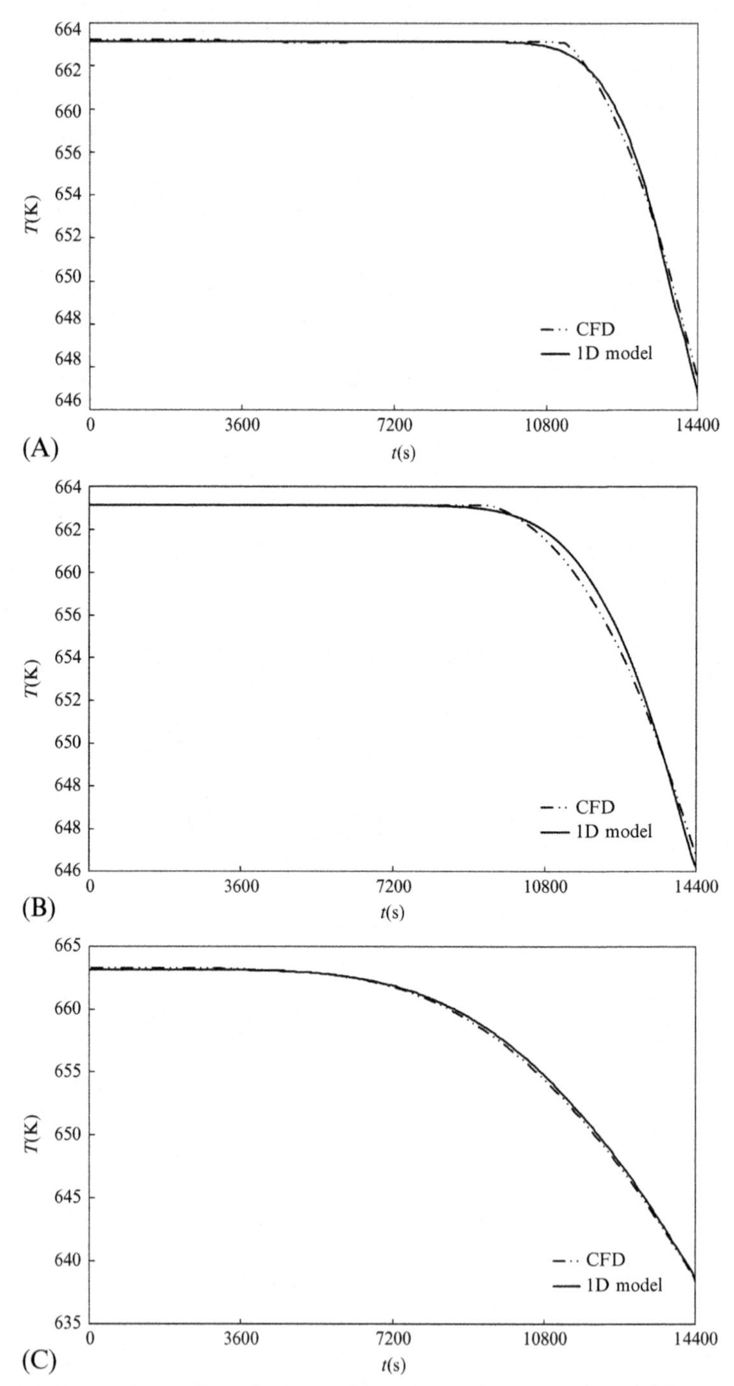

Fig. 4.28 Comparisons of results from CFD and one-dimensional model for the time variation of exit temperature of HTF. (A) Plate case; (B) cylinder case; (C) tube case.

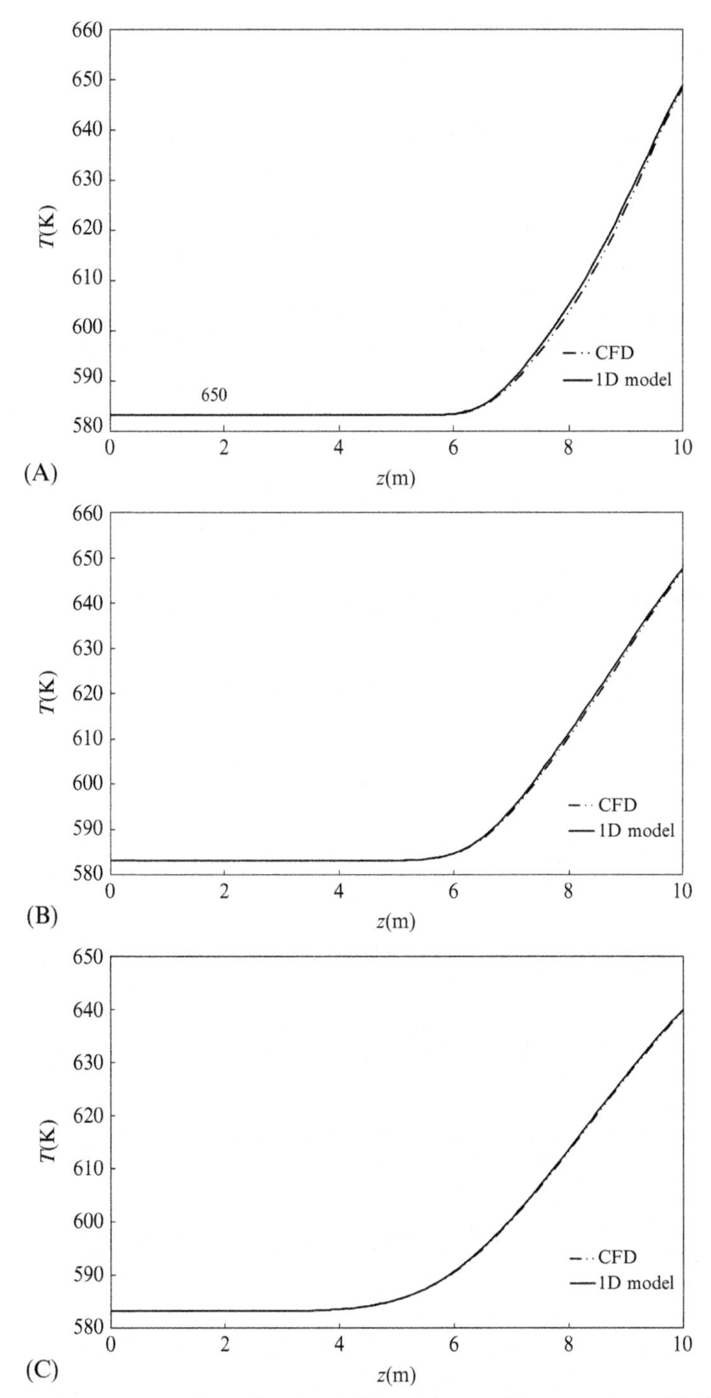

Fig. 4.29 Comparisons of results from CFD and one-dimensional model for the temperature distribution of HTF. (A) Plate case; (B) cylinder case; (C) tube case.

Finally, it is worth noting that, typically, the Schumann equations (1D transient model) for thermal energy storage ignore the axial heat conduction in both the solid and fluid, which is sufficiently accurate for regular dual-material thermal storage systems. However, if the solid and the liquid are highly conductive metals and liquid metals, respectively, which is rarely the case, the validity of the model has to be further examined.

4.4.2 Latent-Heat Thermal Storage

The 1D latent heat thermal model is developed in Section 4.3.1 and its numerical solution using the method of characteristics is presented in Section 4.3.2. We validate this numerical 1D model by comparing the current simulation results to experimental data by Nallusamy et al. [30]. In their experimental work, they used encapsulated spherical capsules of paraffin with melting temperature at 60°C as the PCM, and the HTF is water. The inlet fluid temperature was maintained at 70°C and the mass flow rate was fixed at 2 L/min. Using their experimental conditions and properties, the important parameters were estimated to be:

$$\begin{aligned} &H_{CR}=1.008;\ \ \tau_r=1.0269;\ \ \theta_{f_{inlet}}=1;\\ &\theta_{f_o}=\theta_{s_o}=0;\ \ \theta_{s\_melt}=0.7368;\ \ \theta_{s\_ref}=0;\\ &\eta_{s\_melt}=1;\ \ \frac{C_{s\_s}}{C_{s\_\ell}}=1.1268;\ \ Stf=0.1143;\\ &\Delta t^*=\Delta z^+=0.001; \end{aligned} \tag{4.95}$$

Figs. 4.30 and 4.31 respectively show the temperatures of the fluid and the PCM at the middle of the tank as a function of time during a heat charging process. The results from simulation and experimental tests are in satisfactory agreement, which validates the correctness of the modeling and the computer code.

It is worth noting that the current model does not consider any heat loss from the tank due to the assumption of perfect thermal insulation. This will cause some discrepancy between the simulation and test results. Studies of the influence of heat loss on the temperatures in the thermal storage tanks have been reported by Modi and Pérez-Segarra [46]. The heat loss at the tank surface needs a certain length of time to penetrate and influence the temperature at the center of the tank. Nevertheless, the currently proposed 1D model has the capacity to incorporate the heat loss by adding a heat loss term on the right-hand side of Eq. (4.62), with little to no sacrificing of computational efficiency.

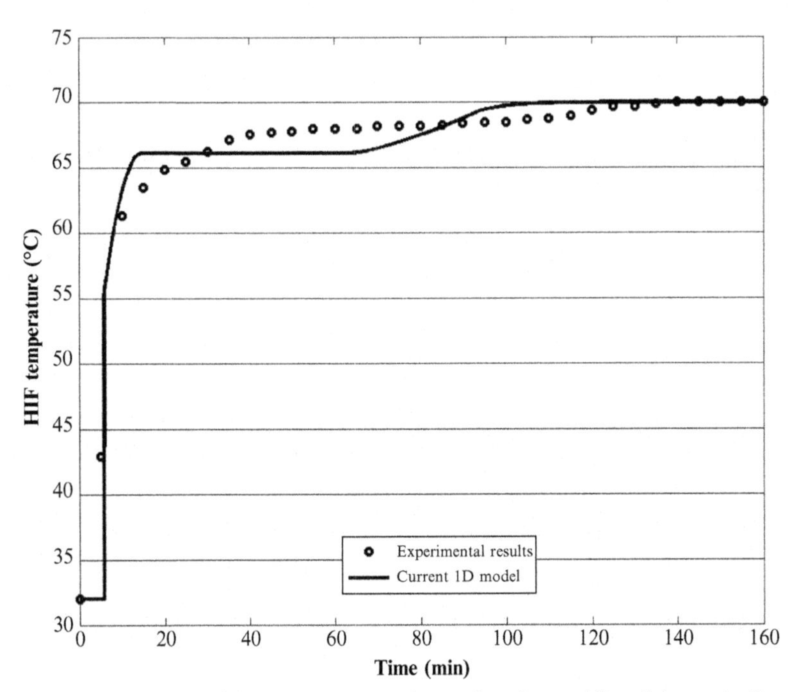

Fig. 4.30 Comparison of fluid temperature located at the middle of the tank. *(Experimental data from Nallusamy N, Sampath S, Velraj R. Experimental investigation on a combined sensible and latent heat storage system integrated with constant/varying (solar) heat sources. Renew Energy 2007;32:1206–27).*

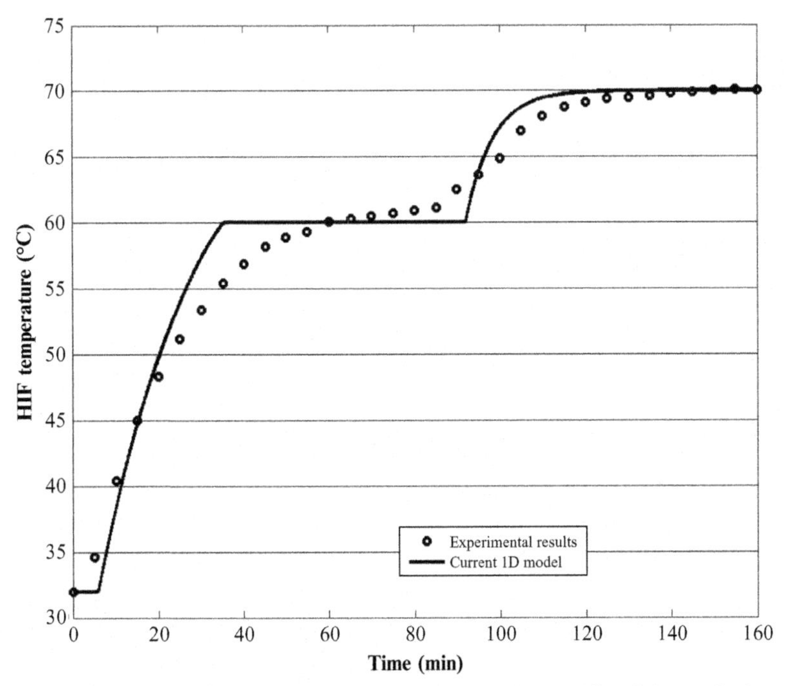

Fig. 4.31 Comparison of PCM temperature located at the middle of the tank. *(Experimental data from Nallusamy N, Sampath S, Velraj R. Experimental investigation on a combined sensible and latent heat storage system integrated with constant/varying (solar) heat sources. Renew Energy 2007;32:1206–27).*

4.5 MODELS DIRECTLY COMPUTE HEAT CONDUCTION IN SOLID AND HTF

The one-dimensional models in Sections 4.2 and 4.3 represent robust, fast, and accurate algorithms. As we have demonstrated, these models can be used for parametric studies and design purposes. When using the effective or corrected heat transfer coefficient in the one-dimensional models to overcome the limit of the lumped heat capacity method, one can deal with all the dual-media thermal storage configurations discussed in Chapter 2. Therefore, the one-dimensional models are cost-effective, convenient, and accurate for design of all types of thermal storage systems.

When detailed local temperature distributions and local heat transfer rates (between a solid material and HTF) are needed, more detailed modeling and algorithms may be employed. However, the computational efforts and time will increase significantly in the modeling if the heat conduction in fluid and solid is computed numerically.

The heat transfer phenomenon in thermal energy storage is truly a conjugate problem. The convection within the HTF and the conduction within the filler materials and container are ongoing simultaneously. These two heat transfer phenomena are coupled and affecting each other. Consequently, they should be solved simultaneously. In the one-dimensional models presented in Sections 4.2 and 4.3, we make use of the heat transfer coefficient and the effective heat transfer coefficient to couple them. This is a very effective way to simplify the conjugate heat transfer problem.

One logical extension of the one-dimensional model is to extend it to two- or three-dimensional formulations. Yang and Garimella [47] introduced an axisymmetric model with a volume-average continuum, momentum, and energy equations for the HTF and lumped capacitance energy equation for the solid filler materials. The coupling between the HTF and filler was treated using an empirical heat transfer coefficient. The equations can be numerically solved. In their work [47] the computations were conducted with the help of the commercial software FLUENT (FLUENT 6.1 Documentation). Flueckiger et al. [48] made use of this axisymmetric model and extended it to include a composite wall with the thermal insulation included in the computational domain. This allowed them to investigate the thermal ratcheting problem in a thermal storage tank.

Xu et al. [49] developed a transient axisymmetric model similar to that of Yang and Garimella. The fluid flow of the HTF is modeled as a porous medium, while the energy conservation is modeled by a convection equation for the HTF and a heat diffusion equation for the solid fillers and the insulation layers and steel wall of the tank. The coupling between the HTF and solid filler materials was treated by use of a heat transfer coefficient, which should come from empirical correlations. At the interface between the HTF and the tank, no slip, same temperature, and energy balance were enforced. The finite volume method was implemented to solve the governing equations. The following discussion briefly introduces the previously mentioned modeling.

Considering the thermal storage system in Fig. 4.32, the packed-bed thermal storage region is viewed as a continuous, homogeneous, and isotropic porous medium. The fluid flow and heat transfer is assumed to be uniform and symmetrical about the axis, and therefore the governing equations for transport within the storage tank are two-dimensional. The flow of molten salt through the packed-bed region is mostly laminar and incompressible, and the properties of the solid fillers are constant. The fluid

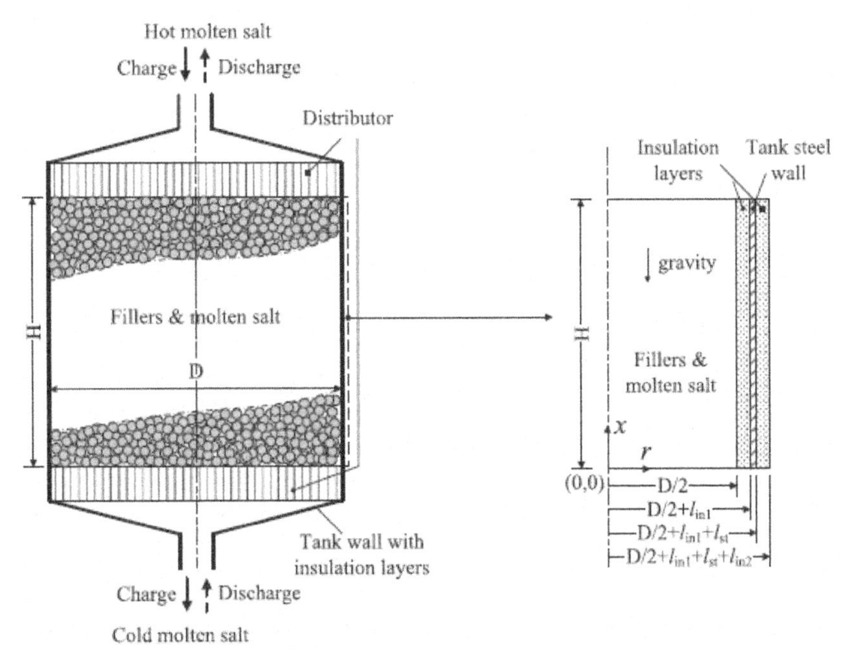

Fig. 4.32 Physical model of packed-bed thermal storage with molten salt HTF. *(Courtesy of Xu C, Wang Z, He Y, Li X, Bai F. Sensitivity analysis of the numerical study on the thermal performance of a packed-bed molten salt thermocline thermal storage system. Appl Energy 2012;92:65-75).*

flow and energy conservation governing equations are established in the following. First, the continuity equation is

$$\frac{\partial\left(\varepsilon\rho_f\right)}{\partial t}+\nabla\cdot\left(\rho_f\vec{u_f}\right)=0 \tag{4.96}$$

and the momentum is

$$\frac{\partial\left(\rho_f\vec{u}_f\right)}{\partial t}+\frac{1}{\varepsilon}\nabla\cdot\left(\rho_f\vec{u}_f\vec{u}_f\right)=-\varepsilon\nabla P+\varepsilon\nabla\cdot\left(\mu_f\nabla\vec{u}_f\right)+\varepsilon\rho_f\vec{g}-\varepsilon\left(\frac{\mu_f}{K}+\frac{C_F\rho_f}{\sqrt{K}}\left|\vec{u}_f\right|\right)\vec{u}_f \tag{4.97}$$

The energy equation for the HTF can be expressed as

$$\frac{\partial\left(\varepsilon\rho_f C_{Pf}T_f\right)}{\partial t}+\nabla\cdot\left(\rho_f C_{Pf}\vec{u}_f T_f\right)=\nabla\cdot\left(k_{eff}\nabla T_f\right)+h\left(T_s-T_f\right) \tag{4.98}$$

Here on the right-hand side, a heat source/sink term, $h(T_s-T_f)$, is introduced to model the heat transfer between the HTF and the solid filler materials. Here h is the volumetric interstitial heat transfer coefficient, which can be found in Ref. [49].

The energy equation for the heat transfer in the solid filler materials is governed by the diffusion equation with a source term:

$$\frac{\partial((1-\varepsilon)\rho_s C_{Ps}T_s)}{\partial t}=\nabla\cdot\left(k_{s,\,eff}\nabla T_s\right)-h\left(T_s-T_f\right) \tag{4.99}$$

To consider the heat transfer between the packed bed and the wall of the storage tank, the governing equation is

$$\rho_w C_{Pw}\frac{\partial(T_w)}{\partial t}=\nabla\cdot\left(k_w\nabla T_w\right) \tag{4.100}$$

Appropriate initial and boundary conditions are needed to solve this set of governing equations. Typically, nonslip and nonpenetration conditions are enforced on the solid boundary. Since sink and source terms are introduced in the energy equation, there is no need of coupling conditions between the HTF and filler material. Along the interface between the HTF and tank wall, temperature continuity and conservation of energy are imposed.

To solve this set of equations, there are a number of possible numerical methods, such as finite volume, finite element, and spectral methods. For convenience, commercial numerical tools (such as ANSYS FLUENT) can be used to solve the governing equations.

APPENDIX STRUCTURE OF THE COMPUTER CODE FOR DUAL-MEDIA THERMAL STORAGE ANALYSIS

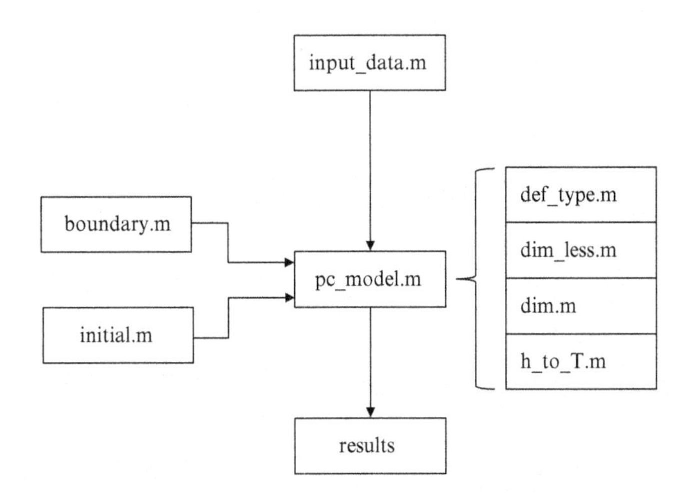

pc_model.m The main code that provides the solution to the governing equations using the method of characteristics, and also conducts postprocess for temperature equilibrium when settled down after each charging/discharging cycle in case multiple cycles of charging/discharging are computed. Results of the computation are also output from the main code.

input_data.m Provides all the required material properties, HTF mass flow rate, charging/discharging time length, size of storage tank, correlation/equation for effective heat transfer coefficient, time step, and mesh size.

boundary.m HTF inlet boundary condition.

initial.m Provides initial temperature of HTF, and initial enthalpy of storage medium (solid or PCM material at solid or liquid state).

def_type.m Definition of melting process, type = 1 is for solid part, type = 2 for melting part, type = 3 for liquid part (for sensible materials, the value of type is always set to be 1).

dim.m Converts dimensionless parameters back to dimensional parameters.

dim_less.m Converts dimensional parameters to dimensionless parameters.

h_to_T.m Equation of state to provide the conversion from enthalpy to temperature for storage media.

Boundary

```
%Phase Change Thermal Energy Storage Model
%Boundary Condition File
%Standard SI Units
```

```
function T_f=boundary(k,t,t_steps,T_f)

for i=1:t_steps+1
     %Boundary Condition for Fluid Temperature

     if rem(k,2)==0

         T_f(i,1)=(310+273.15);

     else

         T_f(i,1)=(390+273.15);

     end

end

end
```

def_type

```
%Phase Change Thermal Energy Storage Model
%Define type for each node

%Standard SI Units
function ans=def_type(h_r,h_melt,stf,k)

%Define type solid
if rem(k,2)==0
   if h_r<=h_melt
      ans=1;

%Define type as melting region
   elseif h_r>h_melt && h_r<=(h_melt+(1/stf))
       ans=2;

%Define type as liquid
   else
       ans=3;
   end
else
     if h_r<=h_melt
        ans=1;

%Define type as melting region
   elseif h_r>h_melt && h_r<(h_melt+(1/stf))
       ans=2;

%Define type as liquid
   else
       ans=3;
     end
end
end
```

dim

```
%Phase Change Thermal Energy Storage Model
%Dimensionalizing File

%Standard SI Units
function out=dim(var,num_ref,den_h,den_l,extra)

out=(var*extra*(den_h-den_l))+num_ref;

end
```

dim_less

```
%Phase Change Thermal Energy Storage Model
%Non-Dimensionalizing File

%Standard SI Units
function [out]=dim_less(var,num_ref,den_h,den_l,extra)

out=(var-num_ref)/(extra*(den_h-den_l));

end
```

h_to_T

```
%Phase Change Thermal Energy Storage Model
%Filler Enthalpy to Temperature Conversion File

%DIMENSIONLESS UNITS!!!!
function ans=h_to_T(h_r,h_r_melt,stf,T_r_melt,T_r_ref,cp_r_s,cp_r_l)

%solid
if h_r<h_r_melt
    ans=h_r*T_r_melt+T_r_ref;

%melting region
elseif h_r>=h_r_melt && h_r<(h_r_melt+(1.0/stf))
    ans=T_r_melt;

%liquid
else %h_r>(h_r_melt+(1/stf))
    ans=((h_r-h_r_melt-1.0/stf)*T_r_melt*cp_r_s/cp_r_l)+T_r_melt;
end

end
```

initial

```
%Phase Change Thermal Energy Storage Model
%Initial Condition File

%Standard SI Units
function [T_f,h_r]=initial(z,z_steps,T_f,h_r,h_r_H,h_r_L,h_r_melt)
```

```
for i=1:z_steps+1
    %Initial Condition for Fluid Temperature
    T_f(1,i)=(26.85+273.15);

    %Initial Condition for filler enthalpy
    h_r(1,i)=h_r_L;

end

end
```

input_data

```
%Phase Change Thermal Energy Storage Model
%Input File

%Standard SI Units
function
[z_o,z_f,del_z,t_o,t_f,del_t,H,A_f,A_r,eps,U,htc,Sr,T_H,T_L,T_r_ref,T_r_melt,
h_r_ref,h_r_melt,h_r_H,h_r_L,L,cp_r_s,cp_r_l,cp_f,rho_f,rho_r]=input_data(k)

%Domain and Step Sizes (DIMENSIONLESS!!!!)
z_o=0;
z_f=1.0;
del_z=1/50;

t_o=0;
if rem(k,2)==0
    t_f=0.0;
else
    t_f=19403.3;
end
del_t=1/50;

%Tank Properties
R=0.1015;                                   %Tank Radius (m)
H=0.6;                                      %Tank Height (m)
A_tank=pi*R*R;                              %Tank Cross Sectional Area (m^2)

eps=0.0407396;                              %Porosity

A_f=eps*A_tank;                             %Effective Fluid Cross Sectional
Area (m^2)
A_r=(1-eps)*A_tank;                         %Effective Filler Cross
Sectional Area (m^2)

%Fluid and Material Properties
m_dot=0.00504854;                           %Inlet mass flow rate (kg/s)

%r=0.02;                                    %Filler Radius (m)
%f_s=2.04;                                  %Surface Factor
Sr=pi*0.0094*19;

T_H=(92+273.15);                            %High Temperature (K)
```

```
T_L=(26.85+273.15);                          %Low Temperature  (K)
T_r_ref=T_L;                                 %Reference Temperature
T_r_melt=(500+273.15);                       %Filler Melting Temperature (K)

cp_r_s=994.893;                              %Filler Solid Specific Heat
(J/kg*K)
cp_r_l=1000;                                 %Filler Liquid Specific Heat
(J/kg*K)

rho_r=2071.43;                               %Filler Density (kg/m^3)

h_r_ref=40000;                               %Filler Reference Enthalpy
(J/kg);
h_r_L=h_r_ref;                               %Filler Enthalpy at T_L
(J/kg);
h_r_melt=cp_r_s*(T_r_melt-T_r_ref)+h_r_ref;  %Filler Melting Enthalpy (J/kg);
L=119000;                                    %Filler Latent Heat (J/kg);
h_r_H=cp_r_s*(T_H-T_L)+h_r_ref;              %Filler Enthalpy at T_H for
sensible case  (J/kg);

cp_f=1005;                                   %Liquid Specific Heat (J/kg*K)

rho_f=1.2585;                                %Fluid Density (kg/m^3)

U=m_dot/rho_f/A_f;                           %Bulk Fluid Velocity (m/s)

kf=0.026;                                    %Fluid conductivity (W/m*K)

ks=0.493;                                    %Solid conductivity (W/m.K)

%nuf=1.807*10^(-5);                          %Fluid Kinematic Viscosity
(m^2/s)

%Pr=nuf*rho_f*cp_f/kf;                       %Prandtl number for fluid

%Re=U*2*R/2/nuf;                             %Reynolds Number for fluid

Nuselt=12.09;                                % Nusselt number

Dtube=0.0094;

htt=Nuselt*kf/(Dtube);                       %Convective Heat Transfer Coef-
ficient (W/(m^2-K))

rii=Dtube/2;

roo=sqrt(R^2/19.0);

B=(rii^3*(4*roo^2-rii^2)+rii*roo^4*(4*log(roo/rii)-3))/(4*(roo^2-rii^2)^2);

htc=1/(1/htt+B/ks);                           %Effective heat transfer coef-
ficient

end
```

pc_model

```
%Phase Change Thermal Energy Storage 1D Model
%Main Function

clc
clear all

for k=1:100% k is used to define the index of cyclic operation

% Enter 'start_type=0' for first run, 'start_type=1' to compute cycles with
% previous ending distributions of temperatures
if k==1
   start_type=0;
end

if k>1
   start_type=1;
end

% Call the subfunction of input_data.m to provide all the required
% parameters for the main function
[z_o,z_f,del_z,t_o,t_f,del_t,H,A_f,A_r,eps,U,htc,Sr,T_H,T_L,T_r_ref,...
T_r_melt,h_r_ref,h_r_melt,h_r_H,h_r_L,L,cp_r_s,cp_r_l,cp_f,rho_f,rho_r]...
=input_data(k);

% Define the dimensionless parameter of tau_r
tau_r=U/H*rho_f*cp_f*A_f/htc/Sr;

% Define the dimensionless paramter of H_CR
H_CR=rho_f*cp_f*eps/rho_r/cp_r_s/(1-eps);

t_steps=round((t_f-t_o)/del_t);    % Total number of time steps
t=(t_o:del_t:t_f);                 % Define t matrix

z_steps=round((z_f-z_o)/del_z);    % Total number of spatial grids
z=(z_o:del_z:z_f);                 % Define z matrix

% Initialization of HTF temp., filler temp. and enthalpy
T_f=zeros(t_steps+1,z_steps+1);
T_r=zeros(t_steps+1,z_steps+1);
h_r=zeros(t_steps+1,z_steps+1);

% Initialization of the filler melting process matrix
type=zeros(t_steps+1,z_steps+1);

% Interfaces Initializations
s=0;                               % Total number of solid interfaces
```

```
l=0;                                % Total number of liquid interfaces
T_f_int=0;                          % Fluid temperature at solid interface
T_f_liq=0;                          % Fluid temperature at liquid interface
h_r_int=0;                          % Filler enthalpy at solid interface
h_r_liq=0;                          % Filler enthalpy at liquid interface
T_r_int=0;                          % Filler temperature at solid interface
T_r_liq=0;                          % Filler temperature at liquid interface

% Index matrix for cases that HTF is traveling too fast
toofast=0;

% Total number of computational errors
error=0;

% Initial value of partial fraction for solid-liquid interface in each grid
beta_new=0;

% Initial value of index matrix for cases that the interface is traveling
% with the same speed with the HTF
test_count=0;

% Initial Condition
if start_type==0 % Initial Run
    cycle_type=1;                   % Equilibrium

    % call the function initial.m
    [T_f,h_r]=initial(z,z_steps,T_f,h_r,h_r_H,h_r_L,h_r_melt);
    s_ct=0;                         % Initialization of the number of solid
interfaces
    l_ct=0;                         % Initialization of the number of liquid
interface

% Cycle computation with results from previous run as initial condition
elseif start_type==1

    % Load the saved data from the previous cycle
    load(['cycle_',num2str(k-1),'.mat'])

    if cycle_type==0
        cycle_type=1;               % Equilibrium
```

```
    else
        cycle_type=1;              % Equilibrium
    end

    T_f(1,:)=T_f_last;             % Initial value of HTF temperature at new
time step
    h_r(1,:)=h_r_last;             % Initial value of filler enthalpy at new
time step
    beta_new=beta_last; % Initial value of Beta at new time step

% Initial value of HTF temp. on the solid and liquid interface at new time
% step
    T_f_int=T_f_int_last;
    T_f_liq=T_f_liq_last;

% Initial value of Filler enthalpy on the solid and liquid interface at new
% time step
    h_r_int=h_r_int_last;
    h_r_liq=h_r_liq_last;

% Initial value of Filler temp. on the solid and liquid interface at new
% time step
    T_r_int=T_r_int_last;
    T_r_liq=T_r_liq_last;

% Initial value of the number of solid and liquid interfaces at new time
% step
    s_ct=s_ct_last;
    l_ct=l_ct_last;
end

% Call the function boundary.m
T_f=boundary(k,t,t_steps,T_f);

% Call the function dim_less.m to calculate the dimensionless HTF
% temperature and the enthalpy of storage medium
for i=1:z_steps+1
    T_f(1,i)=dim_less(T_f(1,i),T_L,T_H,T_L,1);
    h_r(1,i)=dim_less(h_r(1,i),h_r_ref,T_r_melt,T_L,cp_r_s);
end
```

```
%Boundary Condition for Fluid Temperature
for i=2:t_steps+1
    T_f(i,1)=dim_less(T_f(i,1),T_L,T_H,T_L,1);
end

% Calculate the dimensionless melting enthalpy
h_r_melt=dim_less(h_r_melt,h_r_ref,T_r_melt,T_L,cp_r_s);

% Calculate the Stefan number
stf=cp_r_s*(T_r_melt-T_L)/L;

% Calculate the dimensionless reference temperature
T_r_ref=dim_less(T_r_ref,T_L,T_H,T_L,1);

% Calculate the dimensionless melting temperature
T_r_melt=dim_less(T_r_melt,T_L,T_H,T_L,1);

% Create T_r matrix corresponding to h_r based on the equation of state
% definedfunction h_to_T.m
for i=1:z_steps+1
    T_r(1,i)=h_to_T(h_r(1,i),h_r_melt,stf,T_r_melt,T_r_ref,cp_r_s,cp_r_l);
end

% Send the initial value of the number of solid and liquid interface to
% s matrix in the current time step
s=s_ct;
l=l_ct;

%Time iterations for the boundary conditions
for j=1:t_steps
    %Starting from below melting (sensible part)
    if h_r(j,1)<h_r_melt
        type(j,1)=1;
        h_r(j+1,1)=((2*h_r(j,1))-(H_CR*del_t/tau_r)*(h_r(j,1)+(2*T_r_ref...
                    /T_r_melt)-((T_f(j+1,1)+T_f(j,1))/T_r_melt)))...
                    /(2+(H_CR*del_t/tau_r));
        T_r(j+1,1)=h_to_T(h_r(j+1,1),h_r_melt,stf,T_r_melt,T_r_ref,cp_r_s,
cp_r_l);
```

```
        %Check if reached melting (charging process)
        if h_r(j+1,1)>h_r_melt
            disp('new solidus interface 1')
            j
            s=s+1;
            a=del_t*H_CR/tau_r/T_r_melt/2*(T_f(j+1,1)-T_f(j,1));
            b=del_t*H_CR/tau_r/T_r_melt/2*(2*T_f(j,1)-T_r_melt-T_r(j,1));
            c=-1*(h_r_melt-h_r(j,1));

            % Calculate the partial fraction of solid interface at the
            % boundary time iteration
            if a==0
                alpha(s)=-c/b;
            else
                alpha(s)=(-b+sqrt(b^2-4*a*c))/2/a;
            end
            j_melt=j;
            T_f_int(s)=T_f(j,1)+alpha(s)*(T_f(j+1,1)-T_f(j+1,1));
            h_r(j+1,1)=h_r_melt-(1-alpha(s))*del_t*H_CR/tau_r*(1-...
                       ((T_f(j+1,1)+T_f_int(s))/2/T_r_melt));
            T_r(j+1,1)=h_to_T(h_r(j+1,1),h_r_melt,stf,T_r_melt,T_r_ref,
cp_r_s,cp_r_l);

        end

        %Check if breached into liquid region (charging process)
        if h_r(j+1,1)>(h_r_melt+(1/stf))
            disp('new liquidus interface 1')
            l=l+1;
            a=del_t*H_CR/tau_r/T_r_melt/2*(T_f(j+1,1)-T_f_int(s));
            b=del_t*H_CR/tau_r/T_r_melt*(T_f_int(s)-T_r_melt);
            c=-1/stf;

            % Calculate the partial fraction of liquid interface at the
            % boundary time iteration
            if a==0
                gamma(l)=-c/b;
            else
                gamma(l)=(-b+sqrt(b^2-4*a*c))/2/a;
            end
```

```
          j_liq=j;
          T_f_liq(l)=T_f_int(s)+gamma(l)*(T_f(j+1,1)-T_f_int(s));
          h_r(j+1,1)=(2*h_r_melt+2/stf-(H_CR*del_t/tau_r*(1-gamma(l)...
                      -alpha(s)))*((2-(h_r_melt+(1/stf))*(cp_r_s/cp_r_l))...
                      -((T_f(j+1,1)+T_f_liq(l))/T_r_melt)))/(2+(H_CR*...
                      del_t/tau_r*(1-gamma(l)-alpha(s))*cp_r_s/cp_r_l));
          T_r(j+1,1)=h_to_T(h_r(j+1,1),h_r_melt,stf,T_r_melt,T_r_ref,
cp_r_s,cp_r_l);
        end

    %Start in Melting Region
    elseif h_r(j,1)>=h_r_melt && h_r(j,1)<(h_r_melt+(1/stf))
        type(j,1)=2;
        h_r(j+1,1)=h_r(j,1)-del_t*H_CR/tau_r*(1-((T_f(j+1,1)+T_f(j,1))...
                  /2/T_r_melt));
        T_r(j+1,1)=h_to_T(h_r(j+1,1),h_r_melt,stf,T_r_melt,T_r_ref,cp_r_s,
cp_r_l);

        %Check whether breach into liquid region (charging process)
        if h_r(j+1,1)>(h_r_melt+(1/stf))
            disp('new liquidus interface 2')
            j
            l=l+1;
            a=del_t*H_CR/tau_r/T_r_melt/2*(T_f(j+1,1)-T_f(j,1));
            b=del_t*H_CR/tau_r/T_r_melt*(T_f(j,1)-T_r_melt);
            c=-1*(h_r_melt+1/stf-h_r(j,1));

            % Calculate the partial fraction of liquid interface at the
            % boundary time iteration
            if a==0
                gamma(l)=-c/b;
            else
                gamma(l)=(-b+sqrt(b^2-4*a*c))/2/a;
            end

            T_f_liq(l)=T_f(j,1)+gamma(l)*(T_f(j+1,1)-T_f(j,1));
            h_r(j+1,1)=(2*h_r_melt+2/stf-(H_CR*del_t/tau_r*(1-gamma(l))))...
                      *((2-(h_r_melt+(1/stf))*(cp_r_s/cp_r_l))-((T_f(j
+1,1)...
                      +T_f_liq(l))/T_r_melt)))/(2+(H_CR*del_t/tau_r*...
                      (1-gamma(l))*cp_r_s/cp_r_l));
```

```
            T_r(j+1,1)=h_to_T(h_r(j+1,1),h_r_melt,stf,T_r_melt,T_r_ref,
cp_r_s,cp_r_l);
        end

        %Check if back to fully solidified (discharging process)
        if h_r(j+1,1)<h_r_melt
            disp('new solidus interface 2')
            s=s+1;
            a=del_t*H_CR/tau_r/T_r_melt/2*(T_f(j+1,1)-T_f(j,1));
            b=del_t*H_CR/tau_r/T_r_melt/2*(2*T_f(j,1)-T_r_melt-T_r(j,1));
            c=-1*(h_r_melt-h_r(j,1));

            % Calculate the partial fraction of solid interface at the
            % boundary time iteration
            if a==0
                alpha(s)=-c/b;
            else
                alpha(s)=(-b+sqrt(b^2-4*a*c))/2/a;
            end

            T_f_int(s)=T_f(j,1)+alpha(s)*(T_f(j+1,1)-T_f(j,1));
            h_r(j+1,1)=((2*h_r_melt)-((1-alpha(s))*H_CR*del_t/tau_r...
                       /T_r_melt)*(T_r_ref+T_r_melt-T_f(j+1,1)-T_f_int
(s)))...
                       /(2+((1-alpha(s))*H_CR*del_t/tau_r));
            T_r(j+1,1)=h_to_T(h_r(j+1,1),h_r_melt,stf,T_r_melt,T_r_ref,
cp_r_s,cp_r_l);
        end

    %Start in Liquid Region h_r(j,1)>(h_r_melt+(1/stf))
    else
        type(j,1)=3;
        h_r(j+1,1)=(2.0*h_r(j,1)-(H_CR*del_t/tau_r)*...
                  ((1.0-(h_r_melt+(1.0/stf))*(cp_r_s/cp_r_l))+((T_r(j,1)...
                  -T_f(j+1,1)-T_f(j,1))/T_r_melt)))/(2.0+(H_CR*del_t/...
                  tau_r*cp_r_s/cp_r_l));
        T_r(j+1,1)=h_to_T(h_r(j+1,1),h_r_melt,stf,T_r_melt,T_r_ref,cp_r_s,
cp_r_l);

        %Check if return to melting region (discharging process)
        if h_r(j+1,1)<(h_r_melt+(1.0/stf))
```

```
            disp('new liquidus interface 3')
            l=l+1;
            a=del_t*H_CR/tau_r/T_r_melt/2*(T_f(j+1,1)-T_f(j,1));
            b=del_t*H_CR/tau_r/T_r_melt/2*(2*T_f(j,1)-T_r_melt-T_r(j,1));
            c=-(h_r_melt+1/stf-h_r(j,1));

            % Calculate the partial fraction of liquid interface at the
            % boundary time iteration
            if a==0
                gamma(l)=-c/b;
            else
                gamma(l)=(-b+sqrt(b^2-4*a*c))/2/a;
            end
            T_f_liq(l)=T_f(j,1)+gamma(l)*(T_f(j+1,1)-T_f(j,1));
            h_r(j+1,1)=h_r_melt+1/stf-(H_CR*del_t/tau_r*(1-gamma(l)))*...
                      (1.0-((T_f(j+1,1)+T_f_liq(l))/2/T_r_melt));
            T_r(j+1,1)=h_to_T(h_r(j+1,1),h_r_melt,stf,T_r_melt,T_r_ref,
cp_r_s,cp_r_l);
        end

        %Check if fully solidified (discharging process)
        if h_r(j+1,1)<h_r_melt
            disp('new solidus interface 3')
            s=s+1;
            a=del_t*H_CR/tau_r/T_r_melt/2*(T_f(j+1,1)-T_f_liq);
            b=del_t*H_CR/tau_r/T_r_melt/2*(T_f_liq-T_r_melt);
            c=1/stf;

            % Calculate the partial fraction of solid interface at the
            % boundary time iteration
            if a==0
                alpha(s)=-c/b;
            else
                alpha(s)=(-b+sqrt(b^2-4*a*c))/2/a;
            end
            T_f_int(s)=T_f_liq+alpha(s)*(T_f(j+1,1)-T_f_liq(l));
            h_r(j+1,1)=(2*h_r_melt-(H_CR*del_t/tau_r*(1-gamma(l)-alpha(s)))
*...
                      (1+((T_r_ref+T_r_melt-T_f(j+1,1)-T_f_int(s))/
T_r_melt)))/...
                      (2+(H_CR*del_t/tau_r*(1-gamma(l)-alpha(s))));
```

```
            T_r(j+1,1)=h_to_T(h_r(j+1,1),h_r_melt,stf,T_r_melt,T_r_ref,
cp_r_s,cp_r_l);
        end

    end

end

%Define type for remainder of points known
type(t_steps+1,1)=def_type(h_r(t_steps+1,1),h_r_melt,stf,k);
for i=2:z_steps+1
    type(1,i)=def_type(h_r(1,i),h_r_melt,stf,k);
end

% Define the position matrix of solid and liquid interface
pos_int=zeros(s+1,t_steps+1);
track=0;

%'Spatial Sweep' Time Iteration
for j=1:t_steps

    %Reassign Necessary values
    s=s_ct;
    l=l_ct;
    beta_old=beta_new;
    T_f_int_old=T_f_int;
    T_f_liq_old=T_f_liq;
    h_r_int_old=h_r_int;
    h_r_liq_old=h_r_liq;
    T_r_int_old=T_r_int;
    T_r_liq_old=T_r_liq;

    % Display the current time step
    j

    for i=1:z_steps
        if type(j,i)==1
            if type(j,i+1)==1
                % Display('in 1/1'), for solid sensible part;
```

```
            LHS=[(1+del_t/tau_r/2), (-del_t*T_r_melt/2/tau_r);...
                (-del_t*H_CR/2/tau_r/T_r_melt),(1+del_t*H_CR/2/tau_r)];
            RHS=[((1-del_t/tau_r/2)*T_f(j,i)+(del_t/2/tau_r*T_r(j,i))+...
                (del_t/2/tau_r*T_r_ref));((1-del_t*H_CR/tau_r/2)*...
                h_r(j,i+1)+(del_t*H_CR/2/tau_r/T_r_melt*T_f(j,i+1))-...
                (del_t*H_CR/tau_r/T_r_melt*T_r_ref))];
            matrix=LHS\RHS;
            T_f(j+1,i+1)=matrix(1);
            h_r(j+1,i+1)=matrix(2);
            T_r(j+1,i+1)=h_to_T(h_r(j+1,i+1),h_r_melt,stf,T_r_melt,
T_r_ref,cp_r_s,cp_r_l);

        elseif type(j,i+1)==2
            disp('in 1/2'); % Solid interface created
            a=(del_t/tau_r*(T_r(j+1,i)-T_r(j,i)+T_f(j,i)-T_f(j+1,i)))+...
              (del_t/tau_r/(1-beta_old(s+1))*(T_r(j,i+1)+...
              T_f_int_old(s)-T_r_int_old(s)-T_f(j,i+1)))+(2*T_r_melt...
              /H_CR/(1-beta_old(s+1))*(h_r_int_old(s)-h_r(j,i+1)));
            b=(2/(1-beta_old(s+1))*(T_r(j,i+1)-T_r_int_old(s)+...
              T_f_int_old(s)-T_f(j,i+1)))+4*tau_r*T_r_melt/del_t...
              /H_CR/(1-beta_old(s+1))*(h_r_int_old(s)-h_r(j,i+1))...
              +2*(T_f(j+1,i)-T_f(j,i))+del_t/tau_r*(T_r(j,i+1)...
              +T_f(j+1,i)-T_f(j,i+1)-T_r(j+1,i))+del_t/tau_r...
              /(1-beta_old(s+1))*(T_r_int_old(s)+T_f(j,i+1)-T_r(j,i+1)...
              -T_f_int_old(s))+2*T_r_melt/H_CR*(h_r_melt-h_r(j,i+1)-...
              ((h_r_int_old(s)-h_r(j,i+1))/(1-beta_old(s+1))));
            c=4*tau_r*T_r_melt/H_CR/del_t*(h_r_melt-h_r(j,i+1)-...
              ((h_r_int_old(s)-h_r(j,i+1))/(1-beta_old(s+1))))+2*
(T_r_melt...
              +T_r(j,i+1)-T_f(j,i+1)-T_f(j+1,i))+2/(1-beta_old(s+1))...
              *(T_r_int_old(s)+T_f(j,i+1)-T_r(j,i+1)-T_f_int_old(s));

             % Calculate the partial fraction beta_new to track the solid
             % interface
              if a==0
                    disp('a=0')
                    beta_new(s+1)= -c/b;
              else
                    poly_ans=roots([a b c]);
                    beta_new(s+1)= poly_ans(2);
              end
```

```
            % Calculate the temp. of HTF and filler, enthalpy of filler
            % at the solid interface
            T_f_int(s)=T_r_melt+T_r(j,i+1)-T_f(j,i+1)-((1-beta_new(s
+1))...
                       /(1-beta_old(s+1)))*(T_r(j,i+1)-T_r_int_old(s)...
                       +T_f_int_old(s)-T_f(j,i+1))+2*tau_r*T_r_melt...
                       /H_CR/del_t*(h_r_melt-h_r(j,i+1)-((1-beta_new(s
+1))...
                       /(1-beta_old(s+1)))*(h_r_int_old(s)-h_r(j,i+1)));
            h_r_int(s)=h_r_melt;
            T_r_int(s)=h_to_T(h_r_int(s),h_r_melt,stf,T_r_melt,T_r_ref,
cp_r_s,cp_r_l);

            % Track and update the solid interface position
            pos_int(s+1,j)=del_z*(i-1)+beta_old(s+1)*del_z;
            if j==t_steps
                track=track+1;
                last_i(track)=i;
            end

            % Check if new solid liquid interface cross into the
            % next spatial grid
            if beta_new(s+1)>1.0

                % Check whether in last spatial step or not
                if i==z_steps
                   disp('crossed spacial boundary')
                   a= del_t/2/tau_r/beta_old(s+1)*(T_f(j,i)-T_f_int_old
(s))+...
                      del_t/2/tau_r/beta_old(s+1)*(T_r_melt-T_r(j,i));
                   b= del_t/2/tau_r*(T_r(j,i+1)-T_f(j,i+1)+T_f(j,i)-...
                      T_r(j,i))+del_t/2/tau_r/beta_old(s+1)*...
                      (T_f_int_old(s)-T_f(j,i)+T_r(j,i)-T_r_melt)+...
                      (T_f_int_old(s)-T_f(j,i))/beta_old(s+1);
                   c= T_r_melt/H_CR*(h_r_melt-h_r(j,i+1))+T_r_melt...
                      +T_r(j,i+1)-T_f(j,i+1)-T_f(j,i)+(T_f(j,i)-...
                      T_f_int_old(s))/beta_old(s+1);
                   d= 2*tau_r*T_r_melt/H_CR/del_t*(h_r_melt-h_r(j,i+1));
```

```
                        disp('zeta calc')
                        poly_ans_z=roots([a b c d]);
                        for root_find=1:1:3
                            if (poly_ans_z(root_find)>0 && poly_ans_z
(root_find)<1.0)
                                eta=poly_ans_z(root_find);
                                beta_new(s+1)=-1;
                                T_f_int(s)=-1;
                                h_r_int(s)=-1;
                            end
                        end

                        % Calculate HTF temperature at the solid interface
                        T_f_eta=2*tau_r*T_r_melt/H_CR/del_t/eta*...
                                (h_r_melt-h_r(j,i+1)+eta*H_CR*del_t/2/tau_r...
                                /T_r_melt*(T_r_melt+T_r(j,i+1)-T_f(j,i+1)));

                        LHS=[(1+del_t/tau_r/2), -del_t*T_r_melt/tau_r/2;...
                             (-(1-eta)*del_t*H_CR/2/tau_r/T_r_melt), (1+(1-
eta)*H_CR*del_t/tau_r/2)];
                        RHS=[T_f(j,i)+del_t/2/tau_r*(T_r_ref-T_f(j,i)+T_r(j,
i));...
                             h_r_melt-(1-eta)*H_CR*del_t/2/tau_r/T_r_melt*
(T_r_ref+T_r_melt-T_f_eta)];
                        matrix=LHS\RHS;
                        T_f(j+1,i+1)=matrix(1);
                        h_r(j+1,i+1)=matrix(2);
                        T_r(j+1,i+1)=h_to_T(h_r(j+1,i+1),h_r_melt,stf,
T_r_melt,T_r_ref,cp_r_s,cp_r_l);

                        % Calculate the position of solid interface
                        pos_int(s+1,j+1)=del_z*i;

                    else
                        a= 2*(T_r(j,i+2)-T_r(j,i+1)-T_f(j,i+2)+T_f(j,i+1))+...
                           4*tau_r*T_r_melt/H_CR/del_t*(h_r(j,i+1)-h_r(j,i
+2))+...
                           2/beta_old(s+1)*(T_f(j,i)-T_f_int_old(s))+...
                           del_t/tau_r/beta_old(s+1)*(T_r(j,i)-T_r_int_old
(s))+...
```

```
                     del_t/tau_r*(T_r(j,i+2)-T_r(j,i+1)-T_f(j,i+2)+T_f
(j,i+1))+...
                     2*T_r_melt/H_CR*(h_r(j,i+1)-h_r(j,i+2))+...
                     del_t/tau_r/beta_old(s+1)*(T_f_int_old(s)-T_f
(j,i));
                  b= 2*(T_r_melt+T_r(j,i+1)-T_f(j,i+1)-T_f(j,i))+...
                     4*tau_r*T_r_melt/H_CR/del_t*(h_r_melt-h_r(j,i+1))-
...
                     del_t/tau_r*(T_r(j,i)-T_r(j,i+1)+T_f(j,i+1)-T_f(j,
i))+...
                     2*T_r_melt/H_CR*(h_r_melt-h_r(j,i+1));

                  % Check for singulatiry in calculation
                  if (abs(a)<1.0e-10) & (abs(b)<1.0e-10)
                      disp('coefficients are zero')
                      beta_new(s+1)=0;
                  else
                      beta_new(s+1)=-b/a;
                  end

                  % Special Calculation for Beta=0
                  if beta_new(s+1)<=0
                      beta_new(s+1)=0;
                      test_count=test_count+1;
                      test(test_count,:)=[(T_f(j,i)+del_t/2/tau_r...
                          *(T_r_melt+T_r(j,i)-T_f(j,i)))/(1+del_t/2/
tau_r),...
                          (h_r_melt-h_r(j,i+1)+H_CR*del_t/2/tau_r/
T_r_melt...
                          *(T_r_melt+T_r(j,i+1)-T_f(j,i+1)))...
                          /(H_CR*del_t/2/tau_r/T_r_melt),s+1,j];

                      % Calculate the temp. of HTF and filler, enthalpy
of filler
                      % at the solid interface
                      T_f_int(s)=test(test_count,1);
                      T_f(j+1,i+1)=T_f_int(s);
                      h_r(j+1,i+1)=h_r_melt;
                      T_r(j+1,i+1)=h_to_T(h_r(j+1,i+1),h_r_melt,stf,
T_r_melt,T_r_ref,cp_r_s,cp_r_l);

                  else
```

```
                        % Check to make sure HTF is not going faster
                        % than the HTF characteristics
                        if beta_new(s+1)>beta_old(s+1)
                            beta_new(s+1)=beta_old(s+1);
                            disp('going too fast');
                            LHS=[(1+del_t/tau_r/2), -del_t*T_r_melt/tau_r/
2;...
                                (-del_t*H_CR/2/tau_r/T_r_melt),(1
+H_CR*del_t/tau_r/2)];
                            RHS=[T_f_int_old(s)+del_t/2/tau_r*(T_r_ref
+T_r_int_old(s)...
                                -T_f_int_old(s));h_r(j,i+1)+beta_new(s+1)*
(h_r(j,i+2)...
                                -h_r(j,i+1))-H_CR*del_t/2/tau_r/
T_r_melt*...
                                (T_r_ref+T_r(j,i+1)+beta_new(s+1)*(T_r(j,i
+2)...
                                -T_r(j,i+1))-T_f(j,i+1)-beta_new(s+1)*(T_f
(j,i+2)-T_f(j,i+1)))];
                            matrix=LHS\RHS;

                            % Calculate the temp. of HTF and filler,
enthalpy of filler
                            % at the solid interface
                            T_f_int(s)=matrix(1);
                            h_r_int(s)=matrix(2);
                            T_r_int(s)=h_to_T(h_r_int(s),h_r_melt,stf,
T_r_melt,T_r_ref,cp_r_s,cp_r_l);
                            toofast(s+1)=toofast(s+1)+1;
                        else

                            % Calculate the temp. of HTF and filler,
enthalpy of filler
                            % at the solid interface
                            T_f_int(s)=T_r_melt+T_r(j,i+1)-T_f(j,i+1)...
                                        +beta_new(s+1)*(T_r(j,i+2)-...
                                        T_r(j,i+1)-T_f(j,i+2)+...
                                        T_f(j,i+1))+2*tau_r*T_r_melt/...
                                        H_CR/del_t*(h_r_melt-h_r(j,i+1)...
                                        -beta_new(s+1)*(h_r(j,i+2)-h_r(j,i
+1)));
                            h_r_int(s)=h_r_melt;
                            T_r_int(s)=h_to_T(h_r_int(s),h_r_melt,stf,
T_r_melt,T_r_ref,cp_r_s,cp_r_l);
                        end
```

```
                    % Calculate New Values
                    eta=(1-beta_old(s+1))/(beta_new(s+1)-beta_old(s+1)
+1);
                    T_f_eta=T_f_int_old(s)+eta*(T_f_int(s)-T_f_int_old
(s));
                    T_r_eta=T_r_int_old(s)+eta*(T_r_int(s)-T_r_int_old
(s));
                    h_r_eta=h_r_int_old(s)+eta*(h_r_int(s)-h_r_int_old
(s));

                    LHS=[(1+del_t/tau_r/2), -del_t*T_r_melt/tau_r/
2;...
                        (-(1-eta)*del_t*H_CR/2/tau_r/T_r_melt), (1+(1-
eta)*H_CR*del_t/tau_r/2)];
                    RHS=[T_f(j,i)+del_t/2/tau_r*(T_r_ref-T_f(j,i)+T_r
(j,i));...
                        h_r_eta-(1-eta)*H_CR*del_t/2/tau_r/T_r_melt*
(T_r_ref+T_r_eta-T_f_eta)];
                    matrix=LHS\RHS;

                    % Update the temp. and enthalpy of HTF and filler in
                    % the next time step and the following grid (j+1,i
+1)
                    T_f(j+1,i+1)=matrix(1);
                    h_r(j+1,i+1)=matrix(2);
                    T_r(j+1,i+1)=h_to_T(h_r(j+1,i+1),h_r_melt,stf,
T_r_melt,T_r_ref,cp_r_s,cp_r_l);
                end

                if j==t_steps
                    last_i(track)=i+1;
                end
            end

        else % For cases that beta_new < 1.0
            % Calculate New Values
            eta=beta_old(s+1)/(1-beta_new(s+1)+beta_old(s+1));
            T_f_eta=T_f_int_old(s)+eta*(T_f_int(s)-T_f_int_old(s));
            T_r_eta=T_r_int_old(s)+eta*(T_r_int(s)-T_r_int_old(s));
            h_r_eta=h_r_int_old(s)+eta*(h_r_int(s)-h_r_int_old(s));
```

```
                    % Update the temp. and enthalpy of HTF and filler in
                    % the next time step and the following grid (j+1,i+1)
                    T_f(j+1,i+1)=(T_f_eta*(1-(1-eta)*del_t/2/tau_r)+(1-
eta)...
                                *del_t/2/tau_r*(T_r_melt+T_r_eta))/...
                                (1+(1-eta)*del_t/2/tau_r);
                    h_r(j+1,i+1)=h_r(j,i+1)-H_CR*del_t/2/tau_r/T_r_melt*...
                                (T_r_melt+T_r(j,i+1)-T_f(j,i+1)-T_f(j+1,i
+1));
                    T_r(j+1,i+1)=h_to_T(h_r(j+1,i+1),h_r_melt,stf,T_r_melt,
T_r_ref,cp_r_s,cp_r_l);

                end
                %s=s-1;

            else %type(j,i+1)==3
                disp('Heat Transfer is occuring too fast...try a smaller time
step to better resolve');
                error=error+1;
            end

        elseif type(j,i)==2 % Cases for discharging process
            if type(j,i+1)==1
                disp('in 2/1');
                a=(del_t/tau_r*(T_r(j+1,i)-T_r(j,i)+T_f(j,i)-T_f(j+1,i)))+...
                  (del_t/tau_r/(1-beta_old(s+1))*(T_r(j,i+1)+T_f_int_old
(s)...
                  -T_r_int_old(s)-T_f(j,i+1)))+(2*T_r_melt/H_CR/(1-beta_old(s
+1))...
                  *(h_r_int_old(s)-h_r(j,i+1)));
                b=(2/(1-beta_old(s+1))*(T_r(j,i+1)-T_r_int_old(s)+T_f_int_old
(s)...
                  -T_f(j,i+1)))+4*tau_r*T_r_melt/del_t/H_CR/(1-beta_old(s
+1))...
                  *(h_r_int_old(s)-h_r(j,i+1))+2*(T_f(j+1,i)-T_f(j,i))...
                  +del_t/tau_r*(T_r(j,i+1)+T_f(j+1,i)-T_f(j,i+1)-T_r(j+1,i))
+...
                  del_t/tau_r/(1-beta_old(s+1))*(T_r_int_old(s)+T_f(j,i+1)...
                  -T_r(j,i+1)-T_f_int_old(s))+2*T_r_melt/H_CR*(h_r_melt...
                  -h_r(j,i+1)-((h_r_int_old(s)-h_r(j,i+1))/(1-beta_old(s
+1))));
                c=4*tau_r*T_r_melt/H_CR/del_t*(h_r_melt-h_r(j,i+1)-...
                  ((h_r_int_old(s)-h_r(j,i+1))/(1-beta_old(s+1))))+...
```

```
                2*(T_r_melt+T_r(j,i+1)-T_f(j,i+1)-T_f(j+1,i))+...
                2/(1-beta_old(s+1))*(T_r_int_old(s)+T_f(j,i+1)-T_r(j,i+1)-
T_f_int_old(s));

            % Calculate the partial fraction beta_new to track the
            % solid interface for discharging process
             if a==0
                 disp('a=0')
                 beta_new(s+1)= -c/b;
             else
                 poly_ans=roots([a b c]);
                 beta_new(s+1)= poly_ans(2);
             end

             % Calculate the temp. of HTF and filler, enthalpy of filler
             % at the solid interface for discharging process
             T_f_int(s)=T_r_melt+T_r(j,i+1)-T_f(j,i+1)-((1-beta_new(s
+1))...
                       /(1-beta_old(s+1)))*(T_r(j,i+1)-T_r_int_old(s)...
                       +T_f_int_old(s)-T_f(j,i+1))+2*tau_r*T_r_melt...
                       /H_CR/del_t*(h_r_melt-h_r(j,i+1)-((1-beta_new(s
+1))...
                       /(1-beta_old(s+1)))*(h_r_int_old(s)-h_r(j,i+1)));
             h_r_int(s)=h_r_melt;

             T_r_int(s)=h_to_T(h_r_int(s),h_r_melt,stf,T_r_melt,T_r_ref,
cp_r_s,cp_r_l);

             % Track and update the position of solid interface
             pos_int(s+1,j)=del_z*(i-1)+beta_old(s+1)*del_z;
             if j==t_steps
                 track=track+1;
                 last_i(track)=i;
             end

             %Check if new Beta passes into the nex grid
             if beta_new(s+1)>1.0
                %Recalculate Beta New
                 disp('in new z');
```

```
            %Check whether in last spatial grid or not
            if i==z_steps
                disp('crossed spacial boundary')
                a= del_t/2/tau_r/beta_old(s+1)*(T_f(j,i)-T_f_int_old
(s))+...
                    del_t/2/tau_r/beta_old(s+1)*(T_r_melt-T_r(j,i));
                b= del_t/2/tau_r*(T_r(j,i+1)-T_f(j,i+1)+T_f(j,i)-T_r
(j,i))+...
                    del_t/2/tau_r/beta_old(s+1)*(T_f_int_old(s)-...
                    T_f(j,i)+T_r(j,i)-T_r_melt)+...
                    (T_f_int_old(s)-T_f(j,i))/beta_old(s+1);
                c= T_r_melt/H_CR*(h_r_melt-h_r(j,i+1))+...
                    T_r_melt+T_r(j,i+1)-T_f(j,i+1)-T_f(j,i)+...
                    (T_f(j,i)-T_f_int_old(s))/beta_old(s+1);
                d= 2*tau_r*T_r_melt/H_CR/del_t*(h_r_melt-h_r(j,i+1));

                disp('zeta calc')
                poly_ans_z=roots([a b c d]);
                for root_find=1:1:3
                    if (poly_ans_z(root_find)>0 & poly_ans_z
(root_find)<1.0)
                        eta=poly_ans_z(root_find);
                        beta_new(s+1)=-1;
                        T_f_int(s)=-1;
                        h_r_int(s)=-1;
                    end
                end

                T_f_eta=2*tau_r*T_r_melt/H_CR/del_t/eta*...
                        (h_r_melt-h_r(j,i+1)+eta*H_CR*del_t/2/...
                        tau_r/T_r_melt*(T_r_melt+T_r(j,i+1)-T_f(j,i
+1)));
                T_f(j+1,i+1)=(T_f(j,i)+del_t/2/tau_r*(T_r_melt+...
                            T_r(j,i)-T_f(j,i)))/(1+del_t/2/tau_r);
                h_r(j+1,i+1)=h_r_melt-(1-eta)*H_CR*del_t/2/tau_r/
T_r_melt*...
                            (2*T_r_melt-T_f(j+1,i+1)-T_f_eta);
                T_r(j+1,i+1)=h_to_T(h_r(j+1,i+1),h_r_melt,stf,
T_r_melt,T_r_ref,cp_r_s,cp_r_l);

                pos_int(s+1,j+1)=del_z*i;

            else
```

```
                    a=2*(T_r(j,i+2)-T_r(j,i+1)-T_f(j,i+2)+T_f(j,i+1))+...
                        4*tau_r*T_r_melt/H_CR/del_t*(h_r(j,i+1)-h_r(j,i
+2))+...

                        2/beta_old(s+1)*(T_f(j,i)-T_f_int_old(s))+...
                        del_t/tau_r/beta_old(s+1)*(T_r(j,i)-T_r_int_old
(s))+...

                        del_t/tau_r*(T_r(j,i+2)-T_r(j,i+1)-T_f(j,i+2)+T_f
(j,i+1))+...

                        2*T_r_melt/H_CR*(h_r(j,i+1)-h_r(j,i+2))+...
                        del_t/tau_r/beta_old(s+1)*(T_f_int_old(s)-T_f
(j,i));
                    b=2*(T_r_melt+T_r(j,i+1)-T_f(j,i+1)-T_f(j,i))+...
                        4*tau_r*T_r_melt/H_CR/del_t*(h_r_melt-h_r(j,i+1))-
...

                        del_t/tau_r*(T_r(j,i)-T_r(j,i+1)+T_f(j,i+1)-T_f(j,
i))+...

                        2*T_r_melt/H_CR*(h_r_melt-h_r(j,i+1));

                    %Check for singulatiry in calculation
                    if (abs(a)<1.0e-10) & (abs(b)<1.0e-10)
                        disp('coefficients are zero')
                        beta_new(s+1)=0;
                    else
                        beta_new(s+1)=-b/a;
                    end

                    if beta_new(s+1)<=0
                        %Special Calculation for Beta=0
                        beta_new(s+1)=0;
                        test_count=test_count+1;
                        test(test_count,:)=[(T_f(j,i)+del_t/2/tau_r*...
                            (T_r_melt+T_r(j,i)-T_f(j,i)))/(1+del_t/2/
tau_r),...

                            (h_r_melt-h_r(j,i+1)+H_CR*del_t/2/tau_r/
T_r_melt...

                            *(T_r_melt+T_r(j,i+1)-T_f(j,i+1)))...

                            /(H_CR*del_t/2/tau_r/T_r_melt),s+1,j];

                        T_f_int(s)=test(test_count,1);
                        T_f(j+1,i+1)=T_f_int(s);
                        h_r(j+1,i+1)=h_r_melt;
```

```
                    T_r(j+1,i+1)=h_to_T(h_r(j+1,i+1),h_r_melt,stf,
T_r_melt,T_r_ref,cp_r_s,cp_r_l);

                else
                    %Check to make sure not going faster than
                    %the HTF characteristic
                    if beta_new(s+1)>beta_old(s+1)
                        beta_new(s+1)=beta_old(s+1);
                        disp('going too fast');
                        LHS=[(1+del_t/tau_r/2), 0;...
                            (-del_t*H_CR/2/tau_r/T_r_melt), 1];
                        RHS=[T_f_int_old(s)+del_t/2/tau_r*(T_r_melt...
                            +T_r_int_old(s)-T_f_int_old(s));...
                            h_r(j,i+1)+beta_new(s+1)*(h_r(j,i+2)...
                            -h_r(j,i+1))-H_CR*del_t/2/tau_r/
T_r_melt*...
                            (T_r_melt+T_r(j,i+1)+beta_new(s+1)...
                            *(T_r(j,i+2)-T_r(j,i+1))-T_f(j,i+1)-...
                            beta_new(s+1)*(T_f(j,i+2)-T_f(j,i+1)))];
                        matrix=LHS\RHS;
                        T_f_int(s)=matrix(1);
                        h_r_int(s)=matrix(2);
                        T_r_int(s)=h_to_T(h_r_int(s),h_r_melt,stf,
T_r_melt,T_r_ref,cp_r_s,cp_r_l);
                        toofast(s+1)=toofast(s+1)+1;
                    else
                        % Calculate the cases for beta_new<=beta_old
                        T_f_int(s)=T_r_melt+T_r(j,i+1)-T_f(j,i+1)
+beta_new(s+1)*...
                                    (T_r(j,i+2)-T_r(j,i+1)-T_f(j,i
+2)...
                                    +T_f(j,i+1))+2*tau_r*T_r_melt/H_CR/
del_t*...
                                    (h_r_melt-h_r(j,i+1)-beta_new(s+1)*
(h_r(j,i+2)-h_r(j,i+1)));
                        h_r_int(s)=h_r_melt;
                        T_r_int(s)=h_to_T(h_r_int(s),h_r_melt,stf,
T_r_melt,T_r_ref,cp_r_s,cp_r_l);

                    end
```

```
                            %Calculate New Values
                            eta=(1-beta_old(s+1))/(beta_new(s+1)-beta_old(s+1)
+1);
                            T_f_eta=T_f_int_old(s)+eta*(T_f_int(s)-T_f_int_old
(s));
                            T_r_eta=T_r_int_old(s)+eta*(T_r_int(s)-T_r_int_old
(s));
                            h_r_eta=h_r_int_old(s)+eta*(h_r_int(s)-h_r_int_old
(s));

                            % Update the values at (j=1,i=1)
                            T_f(j+1,i+1)=(T_f(j,i)+del_t/2/tau_r*(T_r_melt...
                                         +T_r(j,i)-T_f(j,i)))/(1+del_t/2/
tau_r);
                            h_r(j+1,i+1)=h_r_eta-(1-eta)*H_CR*del_t/2/tau_r/
T_r_melt*...
                                        (T_r_melt+T_r_eta-T_f(j+1,i+1)-
T_f_eta);
                            T_r(j+1,i+1)=h_to_T(h_r(j+1,i+1),h_r_melt,stf,
T_r_melt,T_r_ref,cp_r_s,cp_r_l);
                        end

                        if j==t_steps
                            last_i(track)=i+1;
                        end
                    end

                else
                    % Calculate the cases for beta_new<1.0
                    eta=beta_old(s+1)/(1-beta_new(s+1)+beta_old(s+1));
                    T_f_eta=T_f_int_old(s)+eta*(T_f_int(s)-T_f_int_old(s));
                    T_r_eta=T_r_int_old(s)+eta*(T_r_int(s)-T_r_int_old(s));
                    h_r_eta=h_r_int_old(s)+eta*(h_r_int(s)-h_r_int_old(s));

                    LHS=[(1+(1-eta)*del_t/tau_r/2),-(1-eta)*del_t*T_r_melt/
tau_r/2;...
                         (-del_t*H_CR/2/tau_r/T_r_melt),(1+H_CR*del_t/tau_r/
2)];
                    RHS=[((1-(1-eta)*del_t/tau_r/2)*T_f_eta+(1-eta)*del_t/2/
tau_r*(T_r_eta+T_r_ref));...
                         (h_r(j,i+1)-(del_t*H_CR/2/tau_r/T_r_melt)*(T_r_ref
+T_r(j,i+1)-T_f(j,i+1)))];
                    matrix=LHS\RHS;
                    T_f(j+1,i+1)=matrix(1);
```

```
                    h_r(j+1,i+1)=matrix(2);
                    T_r(j+1,i+1)=h_to_T(h_r(j+1,i+1),h_r_melt,stf,T_r_melt,
T_r_ref,cp_r_s,cp_r_l);
                end
                %s=s-1;

            elseif type(j,i+1)==2
                %disp('in 2/2');
                LHS=[(1+del_t/tau_r/2), 0;...
                     (-del_t*H_CR/2/tau_r/T_r_melt), 1];
                RHS=[((1-del_t/tau_r/2)*T_f(j,i)+(del_t/tau_r*T_r_melt));...
                     (h_r(j,i+1)-(del_t*H_CR/2/tau_r/T_r_melt)*(2*T_r_melt...
                     -T_f(j,i+1)))];
                matrix=LHS\RHS;
                T_f(j+1,i+1)=matrix(1);
                h_r(j+1,i+1)=matrix(2);
                T_r(j+1,i+1)=h_to_T(h_r(j+1,i+1),h_r_melt,stf,T_r_melt,
T_r_ref,cp_r_s,cp_r_l);

            else %type(j,i+1)==3
                disp('in 2/3');

                disp('original beta');
                a=(del_t/tau_r*(T_r(j+1,i)-T_r(j,i)+T_f(j,i)-T_f(j+1,i)))+...
                  (del_t/tau_r/(1-beta_old(s+1))*(T_r(j,i+1)+T_f_liq_old
(1)...
                  -T_r_liq_old(1)-T_f(j,i+1)))+(2*T_r_melt/H_CR/(1-beta_old
(s+1))...
                  *(h_r_liq_old(1)-h_r(j,i+1)));
                b=(2/(1-beta_old(s+1))*(T_r(j,i+1)-T_r_liq_old(1)+T_f_liq_old
(1)-T_f(j,i+1)))+...
                  4*tau_r*T_r_melt/del_t/H_CR/(1-beta_old(s+1))*(h_r_liq_old
(1)-h_r(j,i+1))+...
                  2*(T_f(j+1,i)-T_f(j,i))+del_t/tau_r*(T_r(j,i+1)+T_f(j+1,i)-
T_f(j,i+1)-T_r(j+1,i))+...
                  del_t/tau_r/(1-beta_old(s+1))*(T_r_liq_old(1)+T_f(j,i+1)-
T_r(j,i+1)-T_f_liq_old(1))+...
                  2*T_r_melt/H_CR*(h_r_melt+1.0/stf-h_r(j,i+1)-((h_r_liq_old
(1)...
                  -h_r(j,i+1))/(1-beta_old(s+1))));
                c=4*tau_r*T_r_melt/H_CR/del_t*(h_r_melt+1.0/stf-h_r(j,i+1)-
...
```

```
        ((h_r_liq_old(1)-h_r(j,i+1))/(1-beta_old(s+1))))+...
        2*(T_r_melt+T_r(j,i+1)-T_f(j,i+1)-T_f(j+1,i))+...
        2/(1-beta_old(s+1))*(T_r_liq_old(1)+T_f(j,i+1)-T_r(j,i+1)-
T_f_liq_old(1));
      if a==0
          disp('a=0')
          beta_new(s+1)=-c/b;
      else
          poly_ans=roots([a b c]);
          beta_new(s+1)=poly_ans(2);
      end
      T_f_liq(1)=T_r_melt+T_r(j,i+1)-T_f(j,i+1)-((1-beta_new(s
+1))...
              /(1-beta_old(s+1)))*(T_r(j,i+1)-T_r_liq_old(1)...
              +T_f_liq_old(1)-T_f(j,i+1))+2*tau_r*T_r_melt/H_CR/
del_t*...
              (h_r_melt+1.0/stf-h_r(j,i+1)-((1-beta_new(s+1))...
              /(1-beta_old(s+1)))*(h_r_liq_old(1)-h_r(j,i+1)));
      h_r_liq(1)=h_r_melt+1.0/stf;
      T_r_liq(1)=h_to_T(h_r_liq(1),h_r_melt,stf,T_r_melt,T_r_ref,
cp_r_s,cp_r_l);

      pos_int(s+1,j)=del_z*(i-1)+beta_old(s+1)*del_z;

      if j==t_steps
          track=track+1;
          last_i(track)=i;
      end

      %Check if new Beta cross into the next grid
      if beta_new(s+1)>1.0
          disp('beta > 1')
          %Check whether in last spacial step or not
          if i==z_steps
              disp('crossed spacial boundary')
              a= del_t/2/tau_r/beta_old(s+1)*(T_f(j,i)-T_f_liq_old
(1))+...
                 del_t/2/tau_r/beta_old(s+1)*(T_r_melt-T_r(j,i));
              b= del_t/2/tau_r*(T_r(j,i+1)-T_f(j,i+1)+T_f(j,i)...
                 -T_r(j,i))+del_t/2/tau_r/beta_old(s+1)*(T_f_liq_
old(1)-...
```

```
                        T_f(j,i)+T_r(j,i)-T_r_melt)+(T_f_liq_old(1)-...
                        T_f(j,i))/beta_old(s+1);
                     c= T_r_melt/H_CR*(h_r_melt+1.0/stf-h_r(j,i+1))+...
                        T_r_melt+T_r(j,i+1)-T_f(j,i+1)-T_f(j,i)+...
                        (T_f(j,i)-T_f_liq_old(1))/beta_old(s+1);
                     d= 2*tau_r*T_r_melt/H_CR/del_t*(h_r_melt+1.0/stf-h_r
(j,i+1));

                     disp('zeta calc')
                     poly_ans_z=roots([a b c d]);
                     for root_find=1:1:3
                        if (poly_ans_z(root_find)>0 & poly_ans_z
(root_find)<1.0)
                           eta=poly_ans_z(root_find);
                           beta_new(s+1)=-1;
                           T_f_liq(1)=-1;
                           h_r_liq(1)=-1;
                        end
                     end

                     T_f_eta=2*tau_r*T_r_melt/H_CR/del_t/eta*...
                           (h_r_melt+1.0/stf-h_r(j,i+1)+eta*...
                           H_CR*del_t/2/tau_r/T_r_melt*(T_r_melt+...
                           T_r(j,i+1)-T_f(j,i+1)));
                     T_f(j+1,i+1)=(T_f(j,i)+del_t/2/tau_r*(T_r_melt+...
                              T_r(j,i)-T_f(j,i)))/(1+del_t/2/tau_r);
                     h_r(j+1,i+1)=h_r_melt+1.0/stf-(1-eta)*H_CR*del_t...
                              /2/tau_r/T_r_melt*(2*T_r_melt-T_f(j+1,
i+1)-T_f_eta);
                     T_r(j+1,i+1)=h_to_T(h_r(j+1,i+1),h_r_melt,stf,
T_r_melt,T_r_ref,cp_r_s,cp_r_l);

                     pos_int(s+1,j+1)=del_z*i;

                  else
                     disp('not last block');
                     a=2*(T_r(j,i+2)-T_r(j,i+1)-T_f(j,i+2)+T_f(j,i+1))+...
                        4*tau_r*T_r_melt/H_CR/del_t*(h_r(j,i+1)-h_r(j,
i+2))+...
                        2/beta_old(s+1)*(T_f(j,i)-T_f_liq_old(1))+...
                        del_t/tau_r/beta_old(s+1)*(T_r(j,i)-T_r_liq_old
(1))+...
```

```
                        del_t/tau_r*(T_r(j,i+2)-T_r(j,i+1)-T_f(j,i+2)+T_f
(j,i+1))+...
                        2*T_r_melt/H_CR*(h_r(j,i+1)-h_r(j,i+2))+...
                        del_t/tau_r/beta_old(s+1)*(T_f_liq_old(1)-T_f
(j,i));
                    b=2*(T_r_melt+T_r(j,i+1)-T_f(j,i+1)-T_f(j,i))+...
                        4*tau_r*T_r_melt/H_CR/del_t*(h_r_melt+1/stf-h_r
(j,i+1))-...
                        del_t/tau_r*(T_r(j,i)-T_r(j,i+1)+T_f(j,i+1)-T_f
(j,i))+...
                        2*T_r_melt/H_CR*(h_r_melt+1/stf-h_r(j,i+1));

                    %Check for singulatiry in calculation
                    if (abs(a)<1.0e-10) & (abs(b)<1.0e-10)
                        disp('coefficients are zero')
                        beta_new(s+1)=0;
                    else
                        beta_new(s+1)=-b/a;
                    end

                    %Special Calculation for Beta=0
                    if beta_new(s+1)<=0
                        %disp('beta=0')
                        beta_new(s+1)=1e-4;
                        test_count=test_count+1;
                        test(test_count,:)=[(T_f(j,i)+del_t/2/tau_r*...
                            (T_r_melt+T_r(j,i)-T_f(j,i)))/(1+del_t/2/
tau_r),...
                            (h_r_melt+1/stf-h_r(j,i+1)+H_CR*del_t/2/tau_r/
T_r_melt...
                            *(T_r_melt+T_r(j,i+1)-T_f(j,i+1)))/
(H_CR*del_t/2/tau_r/T_r_melt),...
                            s+1,j];
                        T_f_liq(1)=test(test_count,1);
                        T_f(j+1,i+1)=T_f_liq(1);
                        h_r(j+1,i+1)=h_r_melt+1/stf;
                        T_r(j+1,i+1)=h_to_T(h_r(j+1,i+1),h_r_melt,stf,
T_r_melt,T_r_ref,cp_r_s,cp_r_l);

                    else
```

```
                    %Check to make sure not going faster than
                    %the HTF characteristic
                    if beta_new(s+1)>beta_old(s+1)
                        beta_new(s+1)=beta_old(s+1);
                        disp('going too fast');
                        LHS=[(1+del_t/tau_r/2), 0;...
                            (-del_t*H_CR/2/tau_r/T_r_melt), 1];
                        RHS=[T_f_liq_old(1)+del_t/2/tau_r*(T_r_melt
+T_r_liq_old(1)-T_f_liq_old(1));...
                             h_r(j,i+1)+beta_new(s+1)*(h_r(j,i+2)-h_r
(j,i+1))-H_CR*del_t/2/tau_r/T_r_melt*...
                            (T_r_melt+T_r(j,i+1)+beta_new(s+1)*(T_r
(j,i+2)-T_r(j,i+1))-T_f(j,i+1)-beta_new(s+1)*(T_f(j,i+2)-T_f(j,i+1)))];
                        matrix=LHS\RHS;
                        T_f_liq(1)=matrix(1);
                        h_r_liq(1)=matrix(2);
                        T_r_liq(1)=h_to_T(h_r_liq(1),h_r_melt,stf,
T_r_melt,T_r_ref,cp_r_s,cp_r_l);
                        toofast(s+1)=toofast(s+1)+1;
                    else
                        T_f_liq(1)=T_r_melt+T_r(j,i+1)-T_f(j,i+1)
+beta_new(s+1)*...
                            (T_r(j,i+2)-T_r(j,i+1)-T_f(j,i+2)+T_f
(j,i+1))+...
                            2*tau_r*T_r_melt/H_CR/del_t*...
                            (h_r_melt+1.0/stf-h_r(j,i+1)-beta_new(s+1)
*(h_r(j,i+2)-h_r(j,i+1)));
                        h_r_liq(1)=h_r_melt+1.0/stf;
                        T_r_liq(1)=h_to_T(h_r_liq(1),h_r_melt,stf,
T_r_melt,T_r_ref,cp_r_s,cp_r_l);
                    end

                    %Calculate New Values
                    eta=(1-beta_old(s+1))/(beta_new(s+1)-beta_old(s+1)
+1);
                    T_f_eta=T_f_liq_old(1)+eta*(T_f_liq(1)-T_f_liq_old
(1));
                    T_r_eta=T_r_liq_old(1)+eta*(T_r_liq(1)-T_r_liq_old
(1));
                    h_r_eta=h_r_liq_old(1)+eta*(h_r_liq(1)-h_r_liq_old
(1));

                    T_f(j+1,i+1)=(T_f(j,i)+del_t/2/tau_r*(T_r_melt+T_r
(j,i)-T_f(j,i)))/...
```

```
                                        (1+del_t/2/tau_r);
                            h_r(j+1,i+1)=h_r_eta-(1-eta)*H_CR*del_t/2/tau_r/
T_r_melt*...
                                        (T_r_melt+T_r_eta-T_f(j+1,i+1)-
T_f_eta);
                            T_r(j+1,i+1)=h_to_T(h_r(j+1,i+1),h_r_melt,stf,
T_r_melt,T_r_ref,cp_r_s,cp_r_l);
                        end

                        if j==t_steps
                            last_i(track)=i+1;
                        end
                    end

                else
                    eta=beta_old(s+1)/(1-beta_new(s+1)+beta_old(s+1));
                    T_f_eta=T_f_liq_old(l)+eta*(T_f_liq(l)-T_f_liq_old(l));
                    T_r_eta=T_r_liq_old(l)+eta*(T_r_liq(l)-T_r_liq_old(l));
                    h_r_eta=h_r_liq_old(l)+eta*(h_r_liq(l)-h_r_liq_old(l));

                    LHS=[(1+(1-eta)*del_t/tau_r/2), -cp_r_s/cp_r_l*(1-eta)
*del_t*T_r_melt/tau_r/2;...
                            (-del_t*H_CR/2/tau_r/T_r_melt), (1+H_CR*del_t/tau_r/
2*cp_r_s/cp_r_l)];
                    RHS=[((1-(1-eta)*del_t/tau_r/2)*T_f_eta+(1-eta)*del_t/2/
tau_r*(T_r_melt+T_r_eta-cp_r_s/cp_r_l*T_r_melt*(h_r_melt+1/stf)));...
                            (h_r(j,i+1)-(del_t*H_CR/2/tau_r/T_r_melt)*(T_r_melt
+T_r(j,i+1)-T_f(j,i+1)-cp_r_s/cp_r_l*T_r_melt*(h_r_melt+1/stf)))];
                    matrix=LHS\RHS;
                    T_f(j+1,i+1)=matrix(1);
                    h_r(j+1,i+1)=matrix(2);
                    T_r(j+1,i+1)=h_to_T(h_r(j+1,i+1),h_r_melt,stf,T_r_melt,
T_r_ref,cp_r_s,cp_r_l);
                end
                %l=l-1;
            end

        else %type(j,i)==3

            % This is a special case that we want to avoid by choosing the
            % proper time step
            if type(j,i+1)==1
```

```
        disp('Heat transfer is too fast...try a smaller step size');
        error=error+1;

    elseif type(j,i+1)==2
        disp('in 3/2'); % Liquid interface created

        a=(del_t/tau_r*(T_r(j+1,i)-T_r(j,i)+T_f(j,i)-T_f(j+1,i)))+...
          (del_t/tau_r/(1-beta_old(s+1))*(T_r(j,i+1)+T_f_liq_old(1)-
...
          T_r_liq_old(1)-T_f(j,i+1)))+(2*T_r_melt/H_CR/(1-beta_old
(s+1))...
          *(h_r_liq_old(1)-h_r(j,i+1)));
        b=(2/(1-beta_old(s+1))*(T_r(j,i+1)-T_r_liq_old(1)+...
          T_f_liq_old(1)-T_f(j,i+1)))+4*tau_r*T_r_melt/del_t/...
          H_CR/(1-beta_old(s+1))*(h_r_liq_old(1)-h_r(j,i+1))+...
          2*(T_f(j+1,i)-T_f(j,i))+del_t/tau_r*(T_r(j,i+1)+...
          T_f(j+1,i)-T_f(j,i+1)-T_r(j+1,i))+del_t/tau_r/(1-...
          beta_old(s+1))*(T_r_liq_old(1)+T_f(j,i+1)-T_r(j,i+1)...
          -T_f_liq_old(1))+2*T_r_melt/H_CR*(h_r_melt+1.0/stf-...
          h_r(j,i+1)-((h_r_liq_old(1)-h_r(j,i+1))/(1-beta_old
(s+1))));
        c=4*tau_r*T_r_melt/H_CR/del_t*(h_r_melt+1.0/stf-h_r(j,i+1)...
          -((h_r_liq_old(1)-h_r(j,i+1))/(1-beta_old(s+1))))+2*(...
          T_r_melt+T_r(j,i+1)-T_f(j,i+1)-T_f(j+1,i))+2/(1-beta_old
(s+1))...
          *(T_r_liq_old(1)+T_f(j,i+1)-T_r(j,i+1)-T_f_liq_old(1));

        % Calculate the partial fraction beta_new to track the
        % liquid interface
        if a==0
            disp('a=0')
            beta_new(s+1)=-c/b;
        else
            poly_ans=roots([a b c]);
            beta_new(s+1)=poly_ans(2);
        end

        % Calculate the temp. of HTF and filler, enthalpy of filler
        % at the liquid interface
        T_f_liq(1)=T_r_melt+T_r(j,i+1)-T_f(j,i+1)-((1-beta_new(s+1))/
(1-beta_old(s+1)))*...
```

```
                    (T_r(j,i+1)-T_r_liq_old(1)+T_f_liq_old(1)-T_f(j,i
+1))+...
                    2*tau_r*T_r_melt/H_CR/del_t*...
                    (h_r_melt+1.0/stf-h_r(j,i+1)-((1-beta_new(s+1))...
                    /(1-beta_old(s+1)))*(h_r_liq_old(1)-h_r(j,i+1)));
          h_r_liq(1)=h_r_melt+1.0/stf;
          T_r_liq(1)=h_to_T(h_r_liq(1),h_r_melt,stf,T_r_melt,T_r_ref,
cp_r_s,cp_r_l);

          % Track and update the position of liquid interface
          pos_int(s+1,j)=del_z*(i-1)+beta_old(s+1)*del_z;
          if j==t_steps
              track=track+1;
              last_i(track)=i;
          end

          %Check if new Beta cross into the next grid
          if beta_new(s+1)>1.0
              %Recalculate Beta New
              disp('in new z');

              %Check whether in last spatial grid
              if i==z_steps
                  disp('crossed spatial boundary')
                  a= del_t/2/tau_r/beta_old(s+1)*(T_f(j,i)-T_f_liq_old
(1))+...
                      del_t/2/tau_r/beta_old(s+1)*(T_r_melt-T_r(j,i));
                  b= del_t/2/tau_r*(T_r(j,i+1)-T_f(j,i+1)+T_f(j,i)-T_r
(j,i))+...
                      del_t/2/tau_r/beta_old(s+1)*(T_f_liq_old(1)-T_f
(j,i)+T_r(j,i)-T_r_melt)+...
                      (T_f_liq_old(1)-T_f(j,i))/beta_old(s+1);
                  c= T_r_melt/H_CR*(h_r_melt+1.0/stf-h_r(j,i+1))+...
                      T_r_melt+T_r(j,i+1)-T_f(j,i+1)-T_f(j,i)+(T_f(j,i)-
T_f_liq_old(1))/beta_old(s+1);
                  d= 2*tau_r*T_r_melt/H_CR/del_t*(h_r_melt+1.0/stf-h_r
(j,i+1));

                  disp('zeta calc')
                  poly_ans_z=roots([a b c d]);
                  for root_find=1:1:3
```

```
                        if (poly_ans_z(root_find)>0 & poly_ans_z
(root_find)<1.0)
                            eta=poly_ans_z(root_find);
                            beta_new(s+1)=-1;
                            T_f_liq(1)=-1;
                            h_r_liq(1)=-1;
                        end
                    end

                    T_f_eta=2*tau_r*T_r_melt/H_CR/del_t/eta*...
                            (h_r_melt+1.0/stf-h_r(j,i+1)+eta*H_CR*del_t...
                            /2/tau_r/T_r_melt*(T_r_melt+T_r(j,i+1)-T_f(j,i
+1)));

                    LHS=[(1+del_t/tau_r/2), -del_t*T_r_melt/tau_r...
                        /2*cp_r_s/cp_r_l;(-(1-eta)*del_t*H_CR/2/tau_r...
                        /T_r_melt),(1+(1-eta)*H_CR*del_t/tau_r/2*cp_r_s/
cp_r_l)];
                    RHS=[T_f(j,i)+del_t/2/tau_r*(T_r(j,i)-T_f(j,i)+...
                        T_r_melt-cp_r_s/cp_r_l*T_r_melt*(h_r_melt+1.0/
stf));...
                        h_r_melt+1.0/stf-(1-eta)*H_CR*del_t/2/tau_r/
T_r_melt...
                        *(2*T_r_melt-T_f_eta-T_r_melt*cp_r_s/cp_r_l*
(h_r_melt+1.0/stf))];
                    matrix=LHS\RHS;
                    T_f(j+1,i+1)=matrix(1);
                    h_r(j+1,i+1)=matrix(2);
                    T_r(j+1,i+1)=h_to_T(h_r(j+1,i+1),h_r_melt,stf,
T_r_melt,T_r_ref,cp_r_s,cp_r_l);

                    pos_int(s+1,j+1)=del_z*i;

                else
                    a=2*(T_r(j,i+2)-T_r(j,i+1)-T_f(j,i+2)+T_f(j,i+1))+...
                        4*tau_r*T_r_melt/H_CR/del_t*(h_r(j,i+1)-h_r(j,i
+2))+...
                        2/beta_old(s+1)*(T_f(j,i)-T_f_liq_old(1))+...
                        del_t/tau_r/beta_old(s+1)*(T_r(j,i)-T_r_liq_old
(1))+...
                        del_t/tau_r*(T_r(j,i+2)-T_r(j,i+1)-T_f(j,i+2)+T_f
(j,i+1))+...
```

```
                        2*T_r_melt/H_CR*(h_r(j,i+1)-h_r(j,i+2))+...
                        del_t/tau_r/beta_old(s+1)*(T_f_liq_old(1)-T_f
(j,i));
                    b=2*(T_r_melt+T_r(j,i+1)-T_f(j,i+1)-T_f(j,i))+...
                        4*tau_r*T_r_melt/H_CR/del_t*(h_r_melt+1.0/stf-h_r
(j,i+1))-...
                        del_t/tau_r*(T_r(j,i)-T_r(j,i+1)+T_f(j,i+1)-T_f
(j,i))+...
                        2*T_r_melt/H_CR*(h_r_melt+1.0/stf-h_r(j,i+1));

                    %Check for singulatiry in calculation
                    if (abs(a)<1.0e-10) & (abs(b)<1.0e-10)
                        disp('coefficients are zero')
                        beta_new(s+1)=0;
                    else
                        beta_new(s+1)=-b/a;
                        disp('got here')
                    end

                    %Special Calculation for Beta=0
                    if beta_new(s+1)<=0
                        beta_new(s+1)=1e-4;
                        test_count=test_count+1;
                        test(test_count,:)=[(T_f(j,i)+del_t/2/tau_r*...
                            (T_r_melt+T_r(j,i)-T_f(j,i)))/(1+del_t/2/
tau_r),...
                            (h_r_melt+1/stf-h_r(j,i+1)+H_CR*del_t/2/
tau_r...
                            /T_r_melt*(T_r_melt+T_r(j,i+1)-T_f(j,i+1)))...
                            /(H_CR*del_t/2/tau_r/T_r_melt),s+1,j];

                        T_f_liq(1)=test(test_count,1);
                        T_f(j+1,i+1)=T_f_liq(1);
                        h_r(j+1,i+1)=h_r_melt+1/stf;

                        T_r(j+1,i+1)=h_to_T(h_r(j+1,i+1),h_r_melt,stf,
T_r_melt,T_r_ref,cp_r_s,cp_r_l);

                    else
```

```
                    %Check to make sure not going faster than the
                    %HTF characteristic
                    if beta_new(s+1)>beta_old(s+1)
                        beta_new(s+1)=beta_old(s+1);
                        disp('going too fast')

                        LHS=[(1+del_t/tau_r/2), -del_t*T_r_melt...
                            /2/tau_r*cp_r_s/cp_r_l;...
                            (-del_t*H_CR/2/tau_r/T_r_melt),...
                            (1+H_CR*del_t/2/tau_r*cp_r_s/cp_r_l)];
                        RHS=[T_f_liq_old(1)+del_t/2/tau_r*(...
                            T_r_melt+T_r_liq_old(1)-T_f_liq_old(1)...
                            -cp_r_s/cp_r_l*T_r_melt*(h_r_melt+1.0/
stf));...
                            h_r(j,i+1)+beta_new(s+1)*(h_r(j,i+2)...

                            -h_r(j,i+1))-H_CR*del_t/2/tau_r/
T_r_melt*...
                            (T_r_melt+T_r(j,i+1)+beta_new(s+1)...
                            *(T_r(j,i+2)-T_r(j,i+1))-T_f(j,i+1)...
                            -beta_new(s+1)*(T_f(j,i+2)-T_f(j,i+1))...
                            -cp_r_s/cp_r_l*T_r_melt*(h_r_melt+1.0/
stf))];
                        matrix=LHS\RHS;
                        T_f_liq(1)=matrix(1);
                        h_r_liq(1)=matrix(2);
                        T_r_liq(1)=h_to_T(h_r_liq(1),h_r_melt,stf,
T_r_melt,T_r_ref,cp_r_s,cp_r_l);
                        toofast(s+1)=toofast(s+1)+1;

                    else
                        T_f_liq(1)=T_r_melt+T_r(j,i+1)-T_f(j,i+1)
+beta_new(s+1)*...
                                (T_r(j,i+2)-T_r(j,i+1)-T_f(j,i+2)
+T_f(j,i+1))+...
                                2*tau_r*T_r_melt/H_CR/del_t*...
                                (h_r_melt+1.0/stf-h_r(j,i+1)-beta_
new(s+1)...
                                *(h_r(j,i+2)-h_r(j,i+1)));
                        h_r_liq(1)=h_r_melt+1.0/stf;
                        T_r_liq(1)=h_to_T(h_r_liq(1),h_r_melt,stf,
T_r_melt,T_r_ref,cp_r_s,cp_r_l);
                    end
```

```
                    %Calculate New Values
                    eta=(1-beta_old(s+1))/(beta_new(s+1)-beta_old(s+1)
+1);
                    T_f_eta=T_f_liq_old(1)+eta*(T_f_liq(1)-T_f_liq_old
(1));
                    T_r_eta=T_r_liq_old(1)+eta*(T_r_liq(1)-T_r_liq_old
(1));
                    h_r_eta=h_r_liq_old(1)+eta*(h_r_liq(1)-h_r_liq_old
(1));

                    LHS=[(1+del_t/tau_r/2), -del_t*T_r_melt/tau_r/
2*cp_r_s/cp_r_l;...
                        (-(1-eta)*del_t*H_CR/2/tau_r/T_r_melt),...
                        (1+(1-eta)*H_CR*del_t/tau_r/2*cp_r_s/cp_r_l)];
                    RHS=[T_f(j,i)+del_t/2/tau_r*(T_r(j,i)-T_f(j,i)...
                        +T_r_melt-cp_r_s/cp_r_l*T_r_melt*(h_r_melt
+1.0/stf));...
                        h_r_eta-(1-eta)*H_CR*del_t/2/tau_r/
T_r_melt*...
                        (T_r_melt+T_r_eta-T_f_eta-T_r_melt*cp_r_s/
cp_r_l*(h_r_melt+1.0/stf))];
                    matrix=LHS\RHS;
                    T_f(j+1,i+1)=matrix(1);
                    h_r(j+1,i+1)=matrix(2);

                    T_r(j+1,i+1)=h_to_T(h_r(j+1,i+1),h_r_melt,stf,
T_r_melt,T_r_ref,cp_r_s,cp_r_l);
                end

                if j==t_steps
                    last_i(track)=i+1;
                end
            end

        else
            eta=beta_old(s+1)/(1-beta_new(s+1)+beta_old(s+1));
            T_f_eta=T_f_liq_old(1)+eta*(T_f_liq(1)-T_f_liq_old(1));
            T_r_eta=T_r_liq_old(1)+eta*(T_r_liq(1)-T_r_liq_old(1));
            h_r_eta=h_r_liq_old(1)+eta*(h_r_liq(1)-h_r_liq_old(1));
```

```
                T_f(j+1,i+1)=(T_f_eta*(1-(1-eta)*del_t/2/tau_r)+
(1-eta)...
                        *del_t/2/tau_r*(T_r_melt+T_r_eta))/(1+(1-
eta)*del_t/2/tau_r);
                h_r(j+1,i+1)=h_r(j,i+1)-H_CR*del_t/2/tau_r/T_r_melt*...
                        (T_r_melt+T_r(j,i+1)-T_f(j,i+1)-T_f(j+1,i
+1));
                T_r(j+1,i+1)=h_to_T(h_r(j+1,i+1),h_r_melt,stf,T_r_melt,
T_r_ref,cp_r_s,cp_r_l);

            end
            %l=l-1;

        else %type(j,i+1)==3
            %disp('in 3/3');

            LHS=[(1+del_t/tau_r/2), (-del_t*T_r_melt/2/tau_r*cp_r_s/
cp_r_l);...
                (-del_t*H_CR/2/tau_r/T_r_melt),(1+del_t*H_CR/2/
tau_r*cp_r_s/cp_r_l)];
            RHS=[((1-del_t/tau_r/2)*T_f(j,i)+(del_t/2/tau_r*...
                (T_r(j,i)+T_r_melt-T_r_melt*cp_r_s/cp_r_l*(h_r_melt+1.0/
stf))));...
                (h_r(j,i+1)-(del_t*H_CR/2/tau_r/T_r_melt)*...
                (T_r(j,i+1)+T_r_melt-T_f(j,i+1)-T_r_melt*cp_r_s/cp_r_l*
(h_r_melt+1.0/stf)))];
            matrix=LHS\RHS;
            T_f(j+1,i+1)=matrix(1);
            h_r(j+1,i+1)=matrix(2);
            T_r(j+1,i+1)=h_to_T(h_r(j+1,i+1),h_r_melt,stf,T_r_melt,
T_r_ref,cp_r_s,cp_r_l);
        end

    end

    %Define Type for new calculation
    type(j+1,i+1)=def_type(h_r(j+1,i+1),h_r_melt,stf,k);
end

%Test if Beta calculation is necessary before
%proceeding to next step in time
if type(j,1)==1
    %Onset of Solidous Interface at boundary (heat being added to filler)
```

```
if type(j+1,1)==2
   disp('beta 1/2')
   s_ct=s_ct+1;
   %Initial Beta Calculation
   a=(H_CR*del_t*del_t/tau_r/tau_r/T_r_melt*(T_r(j,2)-T_r(j,1)...
     -T_f(j,2)+T_f(j,1)-T_f(j+1,1)+T_f_int_old(s_ct)+T_r(j+1,1)...
     -T_r_melt)-2*del_t/tau_r*(h_r(j,2)-h_r(j,1)));
   b=(H_CR*del_t*del_t/tau_r/tau_r/T_r_melt*(T_r(j,1)-T_f(j,1)+...
     T_f(j+1,1)-T_r(j+1,1))+2*H_CR*del_t/tau_r/T_r_melt*(T_r(j,2)...
     -T_r(j,1)-T_f(j,2)+T_f(j,1)+T_f(j+1,1)-T_f_int_old(s_ct))...
     -4*(h_r(j,2)-h_r(j,1))+2*del_t/tau_r*(h_r_melt-h_r(j,1)));
   c=2*H_CR*del_t/tau_r/T_r_melt*(T_r_melt+T_r(j,1)-T_f(j,1)...
     -T_f(j+1,1))+4*(h_r_melt-h_r(j,1));
   if a==0
       disp('a=0')
       beta_new(s_ct+l_ct)=-c/b;
   else
       poly_ans=roots([a b c]);
       beta_new(s_ct+l_ct)=poly_ans(2);
   end

   %Check to make sure interface is not travelling faster than
   %the HTF characteristic
   if beta_new(s_ct+l_ct)>(1-alpha(s_ct))
        disp('interface travelling too fast boundary');
        toofast(3)=1;
        beta_new(s_ct+l_ct)=(1-alpha(s_ct));

        LHS=[(1+del_t*(1-alpha(s_ct))/tau_r/2), 0;...
            (-del_t*H_CR/2/tau_r/T_r_melt), 1];
        RHS=[alpha(s_ct)*T_f(j+1,1)+(1-alpha(s_ct))*T_f(j,1)...
            +del_t*(1-alpha(s_ct))/2/tau_r*(2*T_r_melt-...
            alpha(s_ct)*T_f(j+1,1)-(1-alpha(s_ct))*T_f(j,1));...
            alpha(s_ct)*h_r(j,1)+(1-alpha(s_ct))*h_r(j,2)-H_CR*...
            del_t/2/tau_r/T_r_melt*(T_r_melt+alpha(s_ct)*T_r(j,1)...
            +(1-alpha(s_ct))*T_r(j,2)-alpha(s_ct)*T_f(j,1)...
            -(1-alpha(s_ct))*T_f(j,2))];
        matrix=LHS\RHS;
        T_f_int(s_ct)=matrix(1);
        h_r_int(s_ct)=matrix(2);
```

```
            T_r_int(s_ct)=h_to_T(h_r_int(s_ct),h_r_melt,stf,T_r_melt,
T_r_ref,cp_r_s,cp_r_l);

        else
            T_f_int(s_ct)=T_r_melt+T_r(j,1)-T_f(j,1)+...
                    beta_new(s_ct+l_ct)*(T_r(j,2)+T_f(j,1)-T_r(j,1)-
T_f(j,2))+...
                    2*tau_r*T_r_melt/H_CR/del_t*(h_r_melt-h_r(j,1)...
                    -beta_new(s_ct+l_ct)*(h_r(j,2)-h_r(j,1)));
            h_r_int(s_ct)=h_r_melt;
            T_r_int(s_ct)=h_to_T(h_r_int(s_ct),h_r_melt,stf,T_r_melt,
T_r_ref,cp_r_s,cp_r_l);
        end

      elseif type(j+1,1)==3
          disp('Heat Transfer is occuring too fast...try a smaller step size
to better resolve');
      end

    elseif type(j,1)==2

      if type(j+1,1)==3   %Onset of Liquidoous Interface at boundary (heat
being added to filler)
          disp('beta 2/3')
          l_ct=l_ct+1;
        %Initial Beta Calculation
        a=(H_CR*del_t*del_t/tau_r/tau_r/T_r_melt*(T_r(j,2)-T_r(j,1)-...
          T_f(j,2)+T_f(j,1)-T_f(j+1,1)+T_f_liq_old(l_ct)+T_r(j+1,1)...
          -T_r_melt)-2*del_t/tau_r*(h_r(j,2)-h_r(j,1)));
        b=(H_CR*del_t*del_t/tau_r/tau_r/T_r_melt*(T_r(j,1)-T_f(j,1)...
          +T_f(j+1,1)-T_r(j+1,1))+2*H_CR*del_t/tau_r/T_r_melt*...
          (T_r(j,2)-T_r(j,1)-T_f(j,2)+T_f(j,1)+T_f(j+1,1)-...
          T_f_liq_old(l_ct))-4*(h_r(j,2)-h_r(j,1))+2*del_t/tau_r...
          *(h_r_melt+1.0/stf-h_r(j,1)));
        c=2*H_CR*del_t/tau_r/T_r_melt*(T_r_melt+T_r(j,1)-T_f(j,1)-...
          T_f(j+1,1))+4*(h_r_melt+1.0/stf-h_r(j,1));
        if a==0
            disp('a=0')
            beta_new(s_ct+l_ct)=-c/b;
        else
```

```
            poly_ans=roots([a b c]);
            beta_new(s_ct+l_ct)=poly_ans(2);

        end

        %Check to make sure interface is not travelling faster than
        %the HTF characteristic
        if beta_new(s_ct+l_ct)>(1-gamma(l_ct))
            disp('interface travelling too fast');
            toofast(4)=1;
            beta_new(s_ct+l_ct)=(1-gamma(l_ct));

            LHS=[(1+del_t*(1-gamma(l_ct))/tau_r/2), -del_t*T_r_melt...
                *(1-gamma(l_ct))/2/tau_r*cp_r_s/cp_r_l;(-del_t*H_CR...
                /2/tau_r/T_r_melt),1+H_CR*del_t/2/tau_r*cp_r_s/cp_r_l];
            RHS=[gamma(l_ct)*T_f(j+1,1)+(1-gamma(l_ct))*T_f(j,1)+del_t...
                *(1-gamma(l_ct))/2/tau_r*(2*T_r_melt-cp_r_s/cp_r_l...
                *T_r_melt*(h_r_melt+1.0/stf)-gamma(l_ct)*T_f(j+1,1)...
                -(1-gamma(l_ct))*T_f(j,1));gamma(l_ct)*h_r(j,1)...
                +(1-gamma(l_ct))*h_r(j,2)-H_CR*del_t/2/tau_r/T_r_melt*...
                (T_r_melt-cp_r_s/cp_r_l*T_r_melt*(h_r_melt+1.0/stf)...
                +gamma(l_ct)*T_r(j,1)+(1-gamma(l_ct))*T_r(j,2)...
                -gamma(l_ct)*T_f(j,1)-(1-gamma(l_ct))*T_f(j,2))];
            matrix=LHS\RHS;
            T_f_liq(l_ct)=matrix(1);
            h_r_liq(l_ct)=matrix(2);
            T_r_liq(l_ct)=h_to_T(h_r_liq(l_ct),h_r_melt,stf,T_r_melt,
T_r_ref,cp_r_s,cp_r_l);

        else
            T_f_liq(l_ct)=T_r_melt+T_r(j,1)-T_f(j,1)+beta_new(s_ct
+l_ct)...
                            *(T_r(j,2)+T_f(j,1)-T_r(j,1)-T_f(j,2))
+2*tau_r...
                            *T_r_melt/H_CR/del_t*(h_r_melt+1.0/stf-h_r
(j,1)...
                            -beta_new(s_ct+l_ct)*(h_r(j,2)-h_r(j,1)));
            h_r_liq(l_ct)=h_r_melt+1.0/stf;
            T_r_liq(l_ct)=h_to_T(h_r_liq(l_ct),h_r_melt,stf,T_r_melt,
T_r_ref,cp_r_s,cp_r_l);

        end
```

```
        elseif type(j+1,1)==1  %Onset of Solidous Interface at boundary (heat
being removed from filler)
            disp('beta 2/1')
            s_ct=s_ct+1;
            %Beta Calculation
            a=(H_CR*del_t*del_t/tau_r/tau_r/T_r_melt*(T_r(j,2)-T_r(j,1)-...
               T_f(j,2)+T_f(j,1)-T_f(j+1,1)+T_f_int_old(s_ct)+T_r(j+1,1)-...
               T_r_melt)-2*del_t/tau_r*(h_r(j,2)-h_r(j,1)));
            b=(H_CR*del_t*del_t/tau_r/tau_r/T_r_melt*(T_r(j,1)-T_f(j,1)+...
               T_f(j+1,1)-T_r(j+1,1))+2*H_CR*del_t/tau_r/T_r_melt*(T_r
(j,2)...
               -T_r(j,1)-T_f(j,2)+T_f(j,1)+T_f(j+1,1)-T_f_int_old(s_ct))-...
               4*(h_r(j,2)-h_r(j,1))+2*del_t/tau_r*(h_r_melt-h_r(j,1)));
            c= 2*H_CR*del_t/tau_r/T_r_melt*(T_r_melt+T_r(j,1)-T_f(j,1)-...
               T_f(j+1,1))+4*(h_r_melt-h_r(j,1));
            if a==0
               disp('a=0')
               beta_new(s_ct+l_ct)=-c/b;
            else
               poly_ans=roots([a b c]);
               beta_new(s_ct+l_ct)=poly_ans(2);

            end

            %Check to make sure interface is not travelling faster than
            %the HTF characteristic
            if beta_new(s_ct+l_ct)>(1-alpha(s_ct))
                disp('interface travelling too fast');
                beta_new(s_ct+l_ct)=(1-alpha(s_ct));

                LHS=[(1+del_t*(1-alpha(s_ct))/tau_r/2), -del_t*T_r_melt...
                    *(1-alpha(s_ct))/2/tau_r;(-del_t*H_CR/2/tau_r/
T_r_melt)...
                    ,1+H_CR*del_t/2/tau_r];
                RHS=[alpha(s_ct)*T_f(j+1,1)+(1-alpha(s_ct))*T_f(j,1)...
                    +del_t*(1-alpha(s_ct))/2/tau_r*(T_r_melt+T_r_ref...
                    -alpha(s_ct)*T_f(j+1,1)-(1-alpha(s_ct))*T_f(j,1));...
                    alpha(s_ct)*h_r(j,1)+(1-alpha(s_ct))*h_r(j,2)-H_CR*...
                    del_t/2/tau_r/T_r_melt*(T_r_ref+alpha(s_ct)*T_r(j,1)...
```

```
                    +(1-alpha(s_ct))*T_r(j,2)-alpha(s_ct)*T_f(j,1)-...
                    (1-alpha(s_ct))*T_f(j,2))];
                matrix=LHS\RHS;
                T_f_int(s_ct)=matrix(1);
                h_r_int(s_ct)=matrix(2);
                T_r_int(s_ct)=h_to_T(h_r_int(s_ct),h_r_melt,stf,T_r_melt,
T_r_ref,cp_r_s,cp_r_l);

            else
                T_f_int(s_ct)=T_r_melt+T_r(j,1)-T_f(j,1)+beta_new(s_ct
+l_ct)...
                                *(T_r(j,2)+T_f(j,1)-T_r(j,1)-T_f(j,2))
+2*tau_r...
                                *T_r_melt/H_CR/del_t*(h_r_melt-h_r(j,1)...
                                -beta_new(s_ct+l_ct)*(h_r(j,2)-h_r(j,1)));
                h_r_int(s_ct)=h_r_melt;
                T_r_int(s_ct)=h_to_T(h_r_int(s_ct),h_r_melt,stf,T_r_melt,
T_r_ref,cp_r_s,cp_r_l);
            end

        end

    else %type(j,1)==3

%Onset of Liquidoous Interface at boundary (heat being removed from filler)
        if type(j+1,1)==2
           disp('beta 3/2')
           l_ct=l_ct+1;

           %Initial Beta Calculation
           a=(H_CR*del_t*del_t/tau_r/tau_r/T_r_melt*(T_r(j,2)-T_r(j,1)...
             -T_f(j,2)+T_f(j,1)-T_f(j+1,1)+T_f_liq_old(l_ct)+T_r(j+1,1)-
T_r_melt)-...
             2*del_t/tau_r*(h_r(j,2)-h_r(j,1)));
           b=(H_CR*del_t*del_t/tau_r/tau_r/T_r_melt*(T_r(j,1)-T_f(j,1)...
             +T_f(j+1,1)-T_r(j+1,1))+2*H_CR*del_t/tau_r/T_r_melt*(T_r(j,2)-
T_r(j,1)...
             -T_f(j,2)+T_f(j,1)+T_f(j+1,1)-T_f_liq_old(l_ct))-4*(h_r(j,2)-h_r
(j,1))...
             +2*del_t/tau_r*(h_r_melt+1/stf-h_r(j,1)));
```

```
        c=2*H_CR*del_t/tau_r/T_r_melt*(T_r_melt+T_r(j,1)-T_f(j,1)-T_f
(j+1,1))...
          +4*(h_r_melt+1/stf-h_r(j,1));
        if a==0
            disp('a=0')
            beta_new(s_ct+l_ct)=-c/b;
        else
            poly_ans=roots([a b c]);
            beta_new(s_ct+l_ct)=poly_ans(2);
        end

        %Check to make sure interface is not travelling faster than
        %the HTF characteristic
        if beta_new(s_ct+l_ct)>(1-gamma(l_ct))
             disp('interface travelling too fast');
             beta_new(s_ct+l_ct)=(1-gamma(l_ct));

             LHS=[(1+del_t*(1-gamma(l_ct))/tau_r/2), 0;(-del_t*H_CR/2/
tau_r/T_r_melt),1];
             RHS=[gamma(l_ct)*T_f(j+1,1)+(1-gamma(l_ct))*T_f(j,1)...
                 +del_t*(1-gamma(l_ct))/2/tau_r*(2*T_r_melt-gamma(l_ct)...
                 *T_f(j+1,1)-(1-gamma(l_ct))*T_f(j,1));gamma(l_ct)*h_r
(j,1)...
                 +(1-gamma(l_ct))*h_r(j,2)-H_CR*del_t/2/tau_r/T_r_melt...
                 *(T_r_melt+gamma(l_ct)*T_r(j,1)+(1-gamma(l_ct))*T_r(j,2)-
...
                 gamma(l_ct)*T_f(j,1)-(1-gamma(l_ct))*T_f(j,2))];
             matrix=LHS\RHS;
             T_f_liq(l_ct)=matrix(1);
             h_r_liq(l_ct)=matrix(2);
             T_r_liq(l_ct)=h_to_T(h_r_liq(l_ct),h_r_melt,stf,T_r_melt,
T_r_ref,cp_r_s,cp_r_l);

        else
             T_f_liq(l_ct)=T_r_melt+T_r(j,1)-T_f(j,1)+beta_new(s_ct
+l_ct)...
                          *(T_r(j,2)+T_f(j,1)-T_r(j,1)-T_f(j,2))
+2*tau_r...
                          *T_r_melt/H_CR/del_t*(h_r_melt+1.0/stf-h_r
(j,1)...
                          -beta_new(s_ct+l_ct)*(h_r(j,2)-h_r(j,1)));
             h_r_liq(l_ct)=h_r_melt+1.0/stf;
```

```
                T_r_liq(l_ct)=h_to_T(h_r_liq(l_ct),h_r_melt,stf,T_r_melt,
T_r_ref,cp_r_s,cp_r_l);
            end

        elseif type(j+1,1)==1
            disp('Heat Transfer is occuring too fast...try a smaller step size
to better resolve');
        end
    end
end

%Capture Output information for cycling

%Dimensional quantities
T_r_melt=dim(T_r_melt,T_L,T_H,T_L,1);
h_r_melt=dim(h_r_melt,h_r_ref,T_r_melt,T_L,cp_r_s);
T_r_ref=dim(T_r_ref,T_L,T_H,T_L,1);

%Switch coordinates due to cyclic charge and discharge process
for i=1:z_steps+1
     T_f_last(1,z_steps+2-i)=dim(T_f(t_steps+1,i),T_L,T_H,T_L,1);
     h_r_last(1,z_steps+2-i)=dim(h_r(t_steps+1,i),h_r_ref,T_r_melt,T_L,
cp_r_s);
end

%Equilibrium condition after Discharge Process
if cycle_type==1
    disp('apply equil')
    for i=1:z_steps+1
        %Filler at a point that is originally solid
        if h_r_last(1,i)<h_r_melt
            h_r_eq(1,i)=(rho_f*A_f*cp_f*(h_r_ref/cp_r_s-T_r_ref+T_f_last
(1,i))...
                        +rho_r*A_r*(h_r_last(1,i)))/(rho_r*A_r+rho_
f*A_f*cp_f/cp_r_s);
            T_f_eq(1,i)=(h_r_eq(1,i)-h_r_ref)/cp_r_s+T_r_ref;

            if h_r_eq(1,i)>h_r_melt
                T_f_eq(1,i)=(rho_f*A_f*cp_f*T_f_last(1,i)-rho_r*A_r*
(h_r_melt...
                              -h_r_last(1,i)))/(rho_f*A_f*cp_f);
```

```
            h_r_eq(1,i)=(rho_f*A_f*cp_f*(T_f_eq(1,i)-T_r_melt)+rho_r...
                        *A_r*h_r_melt)/(rho_r*A_r);
            T_f_eq(1,i)=T_r_melt;
        end
    end

    %Filler at a point that is originally melting
    if h_r_last(1,i)>=h_r_melt && h_r_last(1,i)<(h_r_melt+L)
       h_r_eq(1,i)=(rho_f*A_f*cp_f*(T_f_last(1,i)-T_r_melt)+rho_r...
                   *A_r*h_r_last(1,i))/(rho_r*A_r);
       T_f_eq(1,i)=T_r_melt;

        if h_r_eq(1,i)>(h_r_melt+L)
           T_f_eq(1,i)=(rho_f*A_f*cp_f*T_f_last(1,i)-rho_r*A_r*
(h_r_melt...
                        +L-h_r_last(1,i)))/(rho_f*A_f*cp_f);
           h_r_eq(1,i)=(rho_f*A_f*cp_f*((h_r_melt+L)/cp_r_l-T_r_melt...
                        +T_f_eq(1,i))+rho_r*A_r*(h_r_melt+L))/(rho_r...
                        *A_r+rho_f*A_f*cp_f/cp_r_l);
           T_f_eq(1,i)=(h_r_eq(1,i)-h_r_melt-L)/cp_r_l+T_r_melt;

        elseif h_r_eq(1,i)<h_r_melt
           T_f_eq(1,i)=(rho_f*A_f*cp_f*T_f_last(1,i)-rho_r*A_r*...
                        (h_r_melt-h_r_last(1,i)))/(rho_f*A_f*cp_f);
           h_r_eq(1,i)=(rho_f*A_f*cp_f*(h_r_ref/cp_r_s-T_r_ref+T_f_eq
(1,i))...
                        +rho_r*A_r*h_r_melt)/(rho_r*A_r+rho_f*A_f*cp_f/
cp_r_s);
           T_f_eq(1,i)=(h_r_eq(1,i)-h_r_ref)/cp_r_s+T_r_ref;
        end
    end

    %Filler at a point that is originally liquid
    if h_r_last(1,i)>=(h_r_melt+L)
           h_r_eq(1,i)=(rho_f*A_f*cp_f*((h_r_melt+L)/cp_r_l-T_r_melt...
                        +T_f_last(1,i))+rho_r*A_r*h_r_last(1,i))...
                        /(rho_r*A_r+rho_f*A_f*cp_f/cp_r_l);
           T_f_eq(1,i)=(h_r_eq(1,i)-h_r_melt-L)/cp_r_l+T_r_melt;

        if h_r_eq(1,i)<(h_r_melt+L)
           T_f_eq(1,i)=(rho_f*A_f*cp_f*T_f_last(1,i)-rho_r*A_r*...
```

```
                        (h_r_melt+L-h_r_last(1,i)))/(rho_f*A_f*cp_f);
                h_r_eq(1,i)=(rho_f*A_f*cp_f*(T_f_eq(1,i)-T_r_melt)+rho_r...
                        *A_r*(h_r_melt+L))/(rho_r*A_r);
                T_f_eq(1,i)=T_r_melt;
            end
        end
    end
    h_r_last=h_r_eq;
    T_f_last=T_f_eq;
end

%Capture necessary interface information
if cycle_type==0
    count=0;
    beta_last=0;
    for i=1:(s_ct+l_ct)
        if beta_new(i)~=-1
            count=count+1;
        end
    end

    for i=1:(s_ct+l_ct)
        if beta_new(i)~=-1
            beta_last(count)=1.0-beta_new(i);
            count=count-1;
        end
    end

    %HTF, filler temperature and enthalpy at the solidus interface
    count=0;
    h_r_int_last=0;
    T_r_int_last=0;
    T_f_int_last=0;
    for i=1:s_ct
        if h_r_int(i)~=-1
            count=count+1;
        end
    end
```

```
    for i=1:s_ct
        if h_r_int(i)~=-1
            h_r_int_last(count)=h_r_int(i);
            T_f_int_last(count)=T_f_int(i);
            T_r_int_last(count)=T_r_int(i);
            count=count-1;
        end
    end

    %HTF, filler temperature and enthalpy at the liquidus interface
    count=0;
    h_r_liq_last=0;
    T_r_liq_last=0;
    T_f_liq_last=0;
    for i=1:l_ct
        if h_r_liq(i)~=-1
            count=count+1;
        end
    end

    for i=1:l_ct
        if h_r_liq(i)~=-1
            h_r_liq_last(count)=h_r_liq(i);
            T_f_liq_last(count)=T_f_liq(i);
            T_r_liq_last(count)=T_r_liq(i);
            count=count-1;
        end
    end
else %equilibrium case
    %Find new interface positions and interface values
    count=0;
    for i=1:(s_ct+l_ct)
        if beta_new(i)~=-1
            count=count+1;
        end
    end

    int_ct=0;
    liq_ct=0;
    beta_last=0;
    h_r_int_last=0;
```

```
    T_f_int_last=0;
    T_r_int_last=0;
    h_r_liq_last=0;
    T_f_liq_last=0;
    T_r_liq_last=0;

    for i=1:z_steps
        %Solidus Interface
        if h_r_last(i)<h_r_melt && h_r_last(i+1)>h_r_melt
            disp('solidus interface created')
            beta_last(count-int_ct-liq_ct)=(h_r_melt-h_r_last(1,i))...
                                                    /(h_r_last(1,i+1)-h_r_last(1,i));
            h_r_int_last(s_ct-int_ct)=dim_less(h_r_melt,h_r_ref,T_r_melt,T_L,
cp_r_s);
            T_f_int_last(s_ct-int_ct)=T_f_last(1,i)+beta_last(count-int_ct-
liq_ct)...
                                              *(T_f_last(1,i+1)-T_f_last(1,i));
            T_f_int_last(s_ct-int_ct)=dim_less(T_r_melt,T_L,T_H,T_L,1);
            T_r_int_last(s_ct-int_ct)=dim_less(T_r_melt,T_L,T_H,T_L,1);
            int_ct=int_ct+1;
        end

        if h_r_last(i)>h_r_melt && h_r_last(i+1)<h_r_melt
            disp('solidus interface created')
            beta_last(count-int_ct-liq_ct)=(h_r_melt-h_r_last(1,i))...
                                                    /(h_r_last(1,i+1)-h_r_last(1,i));
            h_r_int_last(s_ct-int_ct)=dim_less(h_r_melt,h_r_ref,T_r_melt,T_L,
cp_r_s);
            T_f_int_last(s_ct-int_ct)=T_f_last(1,i)+beta_last(count-int_ct-
liq_ct)...
                                              *(T_f_last(1,i+1)-T_f_last(1,i));
            T_f_int_last(s_ct-int_ct)=dim_less(T_r_melt,T_L,T_H,T_L,1);
            T_r_int_last(s_ct-int_ct)=dim_less(T_r_melt,T_L,T_H,T_L,1);
            int_ct=int_ct+1;
        end

        %Liquidus Interface
        if h_r_last(i)<(h_r_melt+L) && h_r_last(i+1)>(h_r_melt+L)
            disp('liquidus interface created')
            beta_last(count-int_ct-liq_ct)=(h_r_melt+L-h_r_last(1,i))...
                                                    /(h_r_last(1,i+1)-h_r_last(1,i));
```

```
            h_r_liq_last(l_ct-liq_ct)=dim_less(h_r_melt,h_r_ref,T_r_melt,T_L,
cp_r_s)+1/stf;
            T_f_liq_last(l_ct-liq_ct)=T_f_last(1,i)+beta_last(count-int_ct-
liq_ct)...
                                       *(T_f_last(1,i+1)-T_f_last(1,i));
            T_f_liq_last(l_ct-liq_ct)=dim_less(T_r_melt,T_L,T_H,T_L,1);
            T_r_liq_last(l_ct-liq_ct)=dim_less(T_r_melt,T_L,T_H,T_L,1);
            liq_ct=liq_ct+1;
         end

          if h_r_last(i)>(h_r_melt+L) && h_r_last(i+1)<(h_r_melt+L)
            disp('liquidus interface created')
            beta_last(count-int_ct-liq_ct)=(h_r_melt+L-h_r_last(1,i))...
                                           /(h_r_last(1,i+1)-h_r_last(1,i));
            h_r_liq_last(l_ct-liq_ct)=dim_less(h_r_melt,h_r_ref,T_r_melt,T_L,
cp_r_s)+1/stf;
            T_f_liq_last(l_ct-liq_ct)=T_f_last(1,i)+beta_last(count-int_ct-
liq_ct)...
                                       *(T_f_last(1,i+1)-T_f_last(1,i));
            T_f_liq_last(l_ct-liq_ct)=dim_less(T_r_melt,T_L,T_H,T_L,1);
            T_r_liq_last(l_ct-liq_ct)=dim_less(T_r_melt,T_L,T_H,T_L,1);
            liq_ct=liq_ct+1;
          end
     end
end

% Update the number of solid and liquid interfaces
s_ct_last=int_ct;
l_ct_last=liq_ct;

% Save the data of current run, use it as the initial data for the next run
save(['cycle_',num2str(k),'.mat'], 'beta_last', 'h_r_last', 'T_f_last',...
    'T_f_int_last', 'h_r_int_last','T_r_int_last', 'T_f_liq_last',...
    'h_r_liq_last','T_r_liq_last','cycle_type','s_ct_last','l_ct_last');
save(['cycle_data_',num2str(k),'.mat'],'T_f','h_r','z','t','-v7.3');

end
```

REFERENCES

[1] Van Lew JT, Li PW, Chan CL, Karaki W, Stephens J. Transient heat delivery and storage process in a thermocline heat storage system, IMECE2009-11701. In: Proceedings of the ASME 2009 international mechanical congress and exposition, November 13–19, Lake Buena Vista, Florida, USA; 2009.
[2] Kays WM, Crawford ME, Weigand B. Convective heat and mass transfer. 4th ed. Boston: McGraw Hill; 2005.
[3] Incropera FP, DeWitt DP. Introduction to heat transfer. 4th ed. New York: John Wiley and Sons, Inc; 2002.
[4] Xu B, Li P-W, Lik Chan C. Extending the validity of lumped capacitance method for large Biot number in thermal storage application. Sol Energy 2012;86(6):1709–24.
[5] Conway JH, Sloane NJH. Sphere packings, lattices and groups. 3rd ed. New York: Springer; 1998. ISBN 0-387-98585-9.
[6] Hales TC. Historical overview of the Kepler conjecture, discrete & computational geometry. Int J Math Comput Sci 2006;36(1):5–20.
[7] Nellis G, Klein S. Heat transfer. New York: Cambridge University Press; 2009.
[8] Bradshaw AV, Johnson A, McLachlan NH, Chiu Y-T. Heat transfer between air and nitrogen and packed beds of non-reacting solids. Trans Instn Chem Eng 1970;48: T77–84.
[9] Jefferson CP. Prediction of breakthrough curves in packed beds: 1. applicability of single parameter models. Am Inst Chem Eng 1972;18(2):409–16.
[10] Hausen H. Wärmeübertragung im Gegenstrom, Gleichstrom und Kreuzstrom. 2nd ed. Berlin: Springer-Verlag; 1976.
[11] Schmidt FW, Willmott AJ. Thermal energy storage and regeneration. New York, NY: McGraw-Hill Book Company; 1981.
[12] Razelos P, Lazaridis A. A lumped heat-transfer coefficient for periodically heated hollow cylinders. Int J Heat Mass Transf 1967;10:1373–87.
[13] Hughes PJ, Klein SA, Close DJ. Packed bed thermal storage models for solar air heating and cooling systems. ASME J Heat Transf 1976;98(2):336–8.
[14] Mumma SA, Marvin WC. A method of simulating the performance of a pebble bed thermal energy storage and recovery system. ASME Paper No. 76-HT-73, In: ASME-AICHE heat transfer conf., St. Louis, Missouri; 1976.
[15] Ozisik MN. Heat Conduction. 2nd ed. New York: Wiley-Interscience; 1993, ISBN 0471532568.
[16] Kakac S. Heat conduction. 3rd ed. Washington, DC: Taylor and Francis; 1993.
[17] Schumann TEW. Heat transfer: a liquid flowing through a porous prism. J Frankl Inst 1929;208(3):405–16.
[18] Shitzer A, Levy M. Transient behavior of a rock-bed thermal storage system subjected to variable inlet air temperatures: analysis and experimentation. J Sol Energy Eng 1983;105(2):200–6.
[19] McMahan AC. Design and optimization of organic Rankine cycle solar-thermal power plants [Master's thesis].Madison, WI: University of Wisconsin; 2006.
[20] Beasley DE, Clark JA. Transient response of a packed bed for thermal energy storage. Int J Heat Mass Transf 1984;27(9):1659–69.
[21] Zarty O, Juddaimi AE. Computational models of a rock-bed thermal storage unit. Sol Wind Technol 1987;2(4):215–8.
[22] McMahan AC, Klein SA, Reindl DT. A finite-time thermodynamic framework for optimizing solar-thermal power plants. J Sol Energy Eng 2007;129(4):355–62.
[23] Pacheco JE, Showalter SK, Kolb WJ. Development of a molten salt thermocline thermal storage system for parabolic trough plants. J Sol Energy Eng 2002;124 (2):153–9.

[24] Kolb GJ, Hassani V. Performance analysis of thermocline energy storage proposed for the 1 mw saguaro solar trough plant. ASME Conf Proc 2006;2006(47454):1–5.
[25] Van Lew JT, Li PW, Chan CL, Karaki W, Stephens J. Analysis of heat storage and delivery of a thermocline tank having solid filler material. J Sol Energy Eng 2011;133:021003.
[26] Courant R, Hilbert D. Methods of mathematical physics, vol. 2. New York: Wiley-Interscience; 1962.
[27] Polyanin AD. Handbook of linear partial differential equations for engineers and scientists. Boca Raton: Chapman & Hall/CRC Press; 2002, ISBN 1-58488-299-9.
[28] Ferziger JH. Numerical methods for engineering applications. New York: Wiley-Interscience; 1998.
[29] Karaki W, Van Lew JT, Li PW, Chan CL, Stephens J. Heat transfer in thermocline storage system with filler materials: analytical model. In: Proceedings of the ASME 2010 4th international conference on energy sustainability, ES2010-90209, May 17–22, Phoenix, Arizona, USA; 2010.
[30] Nallusamy N, Sampath S, Velraj R. Experimental investigation on a combined sensible and latent heat storage system integrated with constant/varying (solar) heat sources. Renew Energy 2007;32:1206–27.
[31] Verma P, Varun, Singal SK. Review of mathematical modeling on latent heat thermal energy storage systems using phase-change material. Renew Sust Energ Rev 2008;12:999–1031.
[32] Tan FL, Hosseinizadeh SF, Khodadadi JM. Liwu Fan, Experimental and computational study of constrained melting of phase change materials (PCM) inside a spherical capsule. Int J Heat Mass Transf 2009;52(15):3464–72.
[33] Zhao W, Elmozughi AF, Oztekien A, Neti S. Heat transfer analysis of encapsulated phase change material for thermal energy storage. Int J Heat Mass Transf 2013;63:323–35.
[34] Regin AF, Solanki SC. An analysis of a packed bed latent heat thermal energy storage system using pcm capsules: numerical investigation. Renew Energy 2009;34:1765–73.
[35] Wu SM, Fang GY, Liu X. Dynamic discharging characteristics simulation on solar heat storage system with spherical capsules using paraffin as heat storage material. Renew Energy 2011;36:1190–5.
[36] Nithyanandam K, Pitchumani R. Thermal energy storage with heat transfer augmentation using thermosyphons. Int J Heat Mass Transf 2013;67:281–94.
[37] Nithyanandam K, Pitchumani R. Computational studies on a latent thermal energy storage system with integral heat pipes for concentrating solar power. Appl Energy 2013;103:400–15.
[38] Archibold AR, Rahman MM, Goswami DY, Stefanakos EL. Parametric investigation of the melting and solidification process in an encapsulated spherical container. In: Proceedings of the ASME 2012 6th international conference on energy sustainability. July 23–26, San Diego, CA, USA; 2012.
[39] Vyshak NR, Jilani G. Numerical analysis of latent heat thermal energy storage system. Energy Convers Manag 2007;48(7):2161–8.
[40] Tumilowicz E, Cho Lik C, Li P, Xu B. An enthalpy formulation for thermocline with encapsulated PCM thermal storage and benchmark solution using the method of characteristics. Int J Heat Mass Transf 2014;79:362–77.
[41] Yang Z, Garimella SV. Molten-salt thermal energy storage in thermoclines under different environmental boundary conditions. Appl Energy 2010;87(11):3322–9.
[42] Li P-W, Van Lew J, Karaki W, Chan C-L, Stephens J, O'Brien JE. Transient heat transfer and energy transport in packed bed thermal storage systems. In: dos Santos Bernardes MA, editor. Developments in Heat Transfer; 2011, ISBN 978-953-307-569-3. http://www.intechopen.com [Accessed September 15].

[43] Becker M. Comparison of heat transfer fluid for use in solar thermal power stations. Electr Power Syst Res 1980;3:139–50.
[44] Li P, Xu B, Han J, Yang Y. Verification of a model of thermal storage incorporated with an extended lumped capacitance method for various solid-fluid structural combinations. Sol Energy 2014;105:71–81.
[45] Li P-W, Van Lew J, Chan C-L, Karaki W, Stephens J, O'Brien JE. Similarity and generalized analysis of efficiencies of thermal energy storage systems. Renew Energy 2012;39:388–402.
[46] Modi A, Pérez-Segarra CD. Thermocline thermal storage systems for concentrated solar power plants: One-dimensional numerical model and comparative analysis. Sol Energy 2014;100:84–93.
[47] Yang Z, Garimella Suresh V. Thermal analysis of solar thermal energy storage in a molten-salt thermocline. Sol Energy 2010;84:974–85.
[48] Flueckiger S, Yang Z, Garimella SV. An Integrated thermal and mechanical investigation of molten-salt thermocline energy storage. Appl Energy 2011;88:2098–105.
[49] Xu C, Wang Z, He Y, Li X, Bai F. Sensitivity analysis of the numerical study on the thermal performance of a packed-bed molten salt thermocline thermal storage system. Appl Energy 2012;92:65–75.

CHAPTER 5

Discharged Fluid Temperatures From a Thermal Storage System—The Key Parameter for Utilization of Stored Thermal Energy

Contents

Abstract

One of the important objectives of the investigation into the behavior of a thermal storage system is to understand the variation of the temperatures of the fluid when discharged from the thermal storage system. It is the discharged fluid temperature that indicates whether the thermal storage system is fully charged or fully discharged in the respective processes. This chapter will discuss the scenarios of the discharged fluid temperature from single medium (HTF alone) and dual-media thermal storage systems under different operating conditions.

Keywords: Dual media thermal storage, Heat transfer fluids, Latent heat thermal storage, Sensible thermal storage, Discharged fluid temperature

Thermal Energy Storage Analyses and Designs
http://dx.doi.org/10.1016/B978-0-12-805344-7.00005-5

5.1 EXAMPLE OF FLUID ALONE SENSIBLE THERMAL STORAGE

When a fluid alone is used for thermal storage, the situation is rather simple: during the energy storage process the fluid pumped out from a cold tank will get heat (from a heat collector) and is then stored in a hot tank; and during the energy discharge process the fluid pumped out from a hot tank will release heat (to a heat sink or thermal energy user) and is then stored in a cold tank. Therefore, for these processes we can assume that the temperature of the discharged fluid from a tank is approximately constant.

5.2 EXAMPLE OF PACKED BEDS BY SOLID FILLER MATERIAL FOR SENSIBLE HEAT STORAGE

The so-called dual media thermal storage system uses a solid packed bed as the primary energy storage material, and a heat transfer fluid flows through the packed bed to deliver or withdraw thermal energy to/from the solid materials. Under this circumstance, the heat transfer between the packed-bed solid material and the heat transfer fluid makes the temperature of the discharged fluid vary during the entire energy charge or discharge process.

Having an accurate prediction of the time-dependent temperature of the fluid flowing out from the dual-media thermal storage tank is critical to the size design of thermal storage tanks and operation of a thermal power plant utilizing the stored thermal energy. For example, when below a certain temperature, the discharged fluid will not be acceptable by the power plant. Therefore, the size design for a thermal storage system needs to consider this requirement.

Fig. 5.1 shows schematically the time-dependent variation of the fluid temperature flowing in and out from a dual-media thermocline sensible thermal storage system. Fig. 5.1A is for a thermal storage/charging process and Fig. 5.1B is for a thermal discharging process [1].

For the thermal discharging process, it is easy to understand that theoretically the temperature of the discharged hot fluid will start to decrease once the original hot fluid in the voids of the packed bed moves out. If the time to fill up a tank volume with fluid is t_{full} and the void fraction of the packed bed is ε, then the time instance that the temperature of the discharged hot fluid

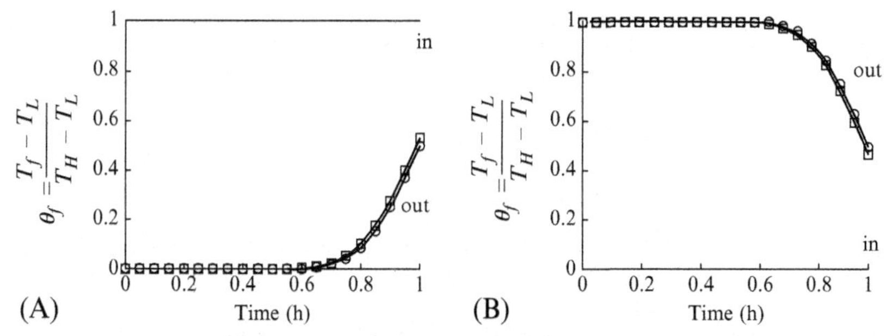

Fig. 5.1 Temperature of fluid at inlet and outlet of thermal storage tank during charging and discharging processes. (A) Thermal charging, (B) thermal discharging.

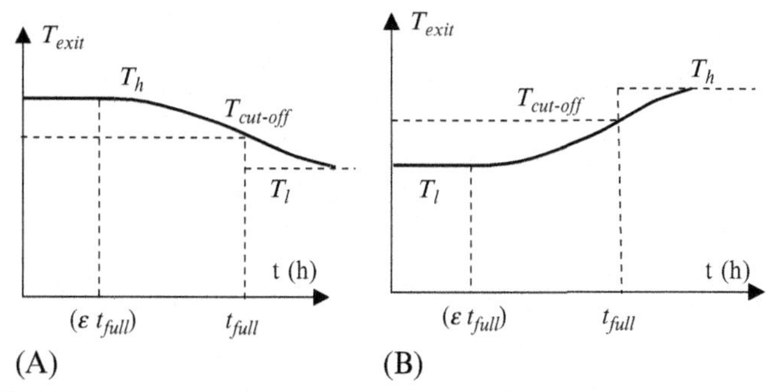

Fig. 5.2 Illustration of the time-dependent outflow fluid temperatures in discharging and charging processes of a dual-media sensible thermal storage system [2]. (A) Discharging, (B) charging.

starts to decrease is $\varepsilon \cdot t_{full}$ as shown in Fig. 5.2A. Similarly, as shown in Fig. 5.2B, in a thermal charging process, the temperature of the cold fluid flowing out of the tank will start to increase at the moment of $\varepsilon \cdot t_{full}$ at which the original cold fluid in the voids of the packed bed moves out.

In the discharging process, a cut-off temperature, as shown in Fig. 5.2A, is typically set as the lowest fluid temperature that is acceptable by the user of the thermal energy. The discharging process stops once the temperature of the fluid out from the tank is below the cut-off temperature. Similarly, it is also important to observe the temperature of the flowing-out fluid in a charging process to control the operation of the thermal charging process. When the flowing-out fluid has a sufficiently high temperature, $T_{cut-off}$, it becomes difficult to charge more energy into the system due to the minimized temperature difference between the solid and the fluid, and the charging process stops.

5.2.1 Multiple Energy Charges and Discharges to Approach Cyclic Steady-State Operation

When a thermal storage system is subjected to operation, the thermal storage tank is initially cold. The thermal charge and discharge operation will repeat for multiple times or days (if one charging and one discharging are made in one day's operation) and then the system comes to a so-called cyclic steady-state operation, if the fluid temperature respectively fed to the charging and discharging processes do not change from one to another time. Typically, after several to 10 charging/discharging cycles, the system will reach a cyclic steady-state. Fig. 5.3 schematically shows the evolution of the dimensionless temperature distribution of the filler material in a tank at the end of each charging and discharging in total six runs of energy storage and withdrawn

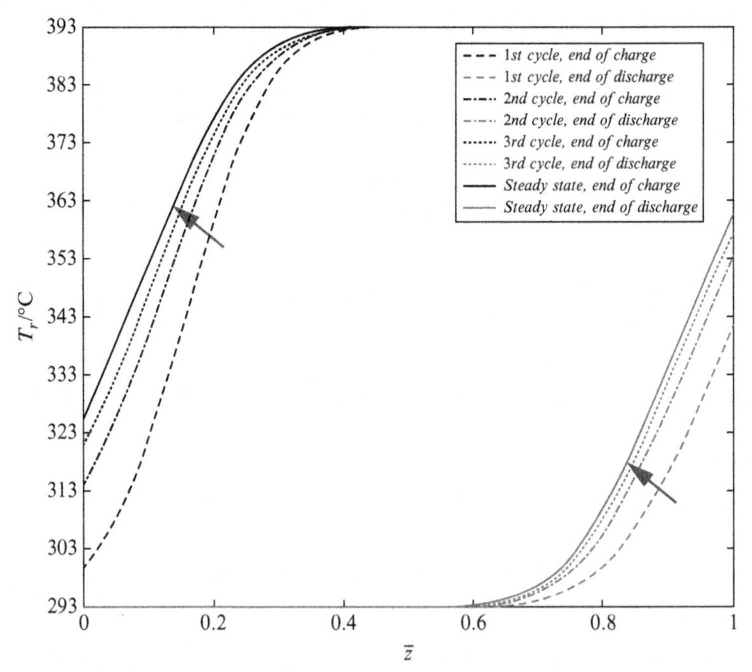

Fig. 5.3 The filler material temperature distribution in a sensible thermal storage tank in six runs of thermal charging and discharging with the tank initially cold [3].

operation. One can see that after three runs of charging/discharging, the temperature distribution in the tank after a charging and a discharging becomes the same as that of the next operation.

5.3 PACKED-BED BY ENCAPSULATED PCM FOR LATENT HEAT THERMAL STORAGE

PCM-based latent heat thermal storage is characterized by a section of constant temperature of fluid flowing out of the storage tank, which is due to the large heat capacity of PCM at the melting or freeze point. Fig. 5.4 shows a typical scenario of the fluid temperature flowing out from a PCM storage tank during a heat discharging process. The first short period shows a quick temperature drop, which is due to the sensible heat in the PCM above the freeze point. The section with a long period of constant temperature is related to the freezing temperature where the release of large latent heat keeps the temperature of fluid constant for a relatively long time. After the section of freezing process, the sharp drop of the temperature of the outflow fluid is due to the relatively low sensible heat capacity of the PCM.

Obviously, if we view the large latent heat as an equivalent of a large heat capacity at the freeze point, we may pack multiple sections of different PCMs with different freeze points in one storage tank, which will allow us to have the advantage of multiple small steps of temperature drop and overall to have small storage volume.

5.3.1 Multiple Energy Charge and Discharges to Approach Cyclic Steady-State Operation for Latent Heat Thermal Storage

In a similar way to that of sensible thermal energy storage, thermal charge and discharge operation for a latent heat thermal storage system will also repeat for multiple times or days (if one charge and one discharge are made in one day's operation) until the system comes to a so-called cyclic steady-state operation. This is assumed that the fluid temperatures respectively fed to the charging and discharging processes do not change each time. Fig. 5.5 schematically shows the evolution of the dimensionless temperature distribution of the filler material in a tank at the end of each charging and discharging in total six runs of energy storage and withdrawn operation. One can

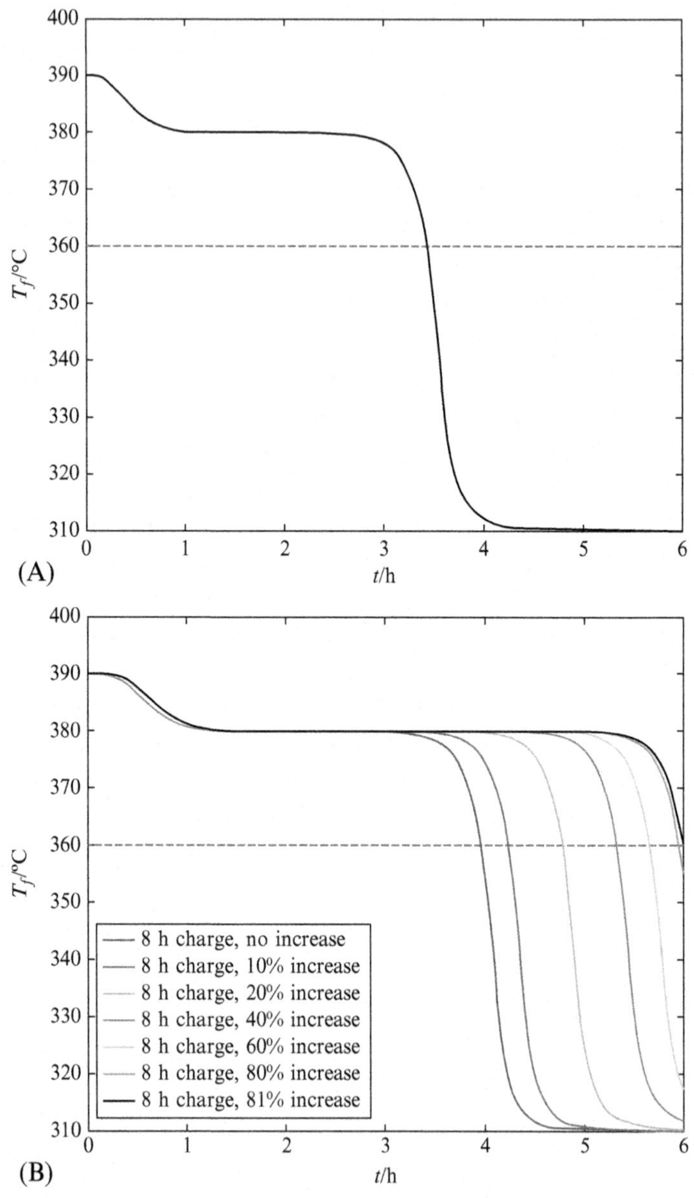

Fig. 5.4 Examples of the temperature of outflow fluid during a discharging process in a PCM thermal storage system [4] (A) A basic case of same charging and discharging time, (B) a case with longer charging time and increased volume on the basic case.

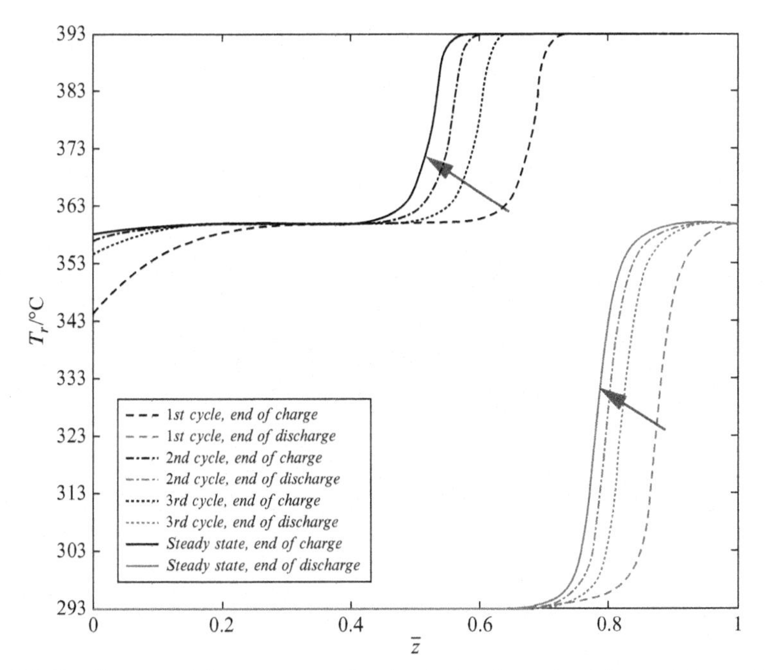

Fig. 5.5 The filler material temperature distribution in a latent heat thermal storage tank in six runs of thermal charging and discharging with the tank initially cold [3].

see that after three runs of charging/discharging, the temperature distribution in the tank becomes the same as that of the next operation.

5.4 MULTIPLE LAYERS OF PACKED BED WITH DIFFERENT PCM OF DIFFERENT PHASE CHANGE TEMPERATURES

Because of the great difference between latent heat and sensible heat, one can see a quick drop of the discharged heat transfer fluid (HTF) once the phase change is complete. For the system to rely more on latent heat thermal storage, the idea of multiple PCM sections with a cascade melting point in one tank has been proposed [5,6], such as is shown in Fig. 5.6. Because each zone in the packed bed has a different PCM material with different melting points, the discharged fluid temperature profile can be different from that of using single PCM. Fig. 5.7 shows the scenario of the fluid temperature measured at the top of PCM-1, PCM-2 and PCM-3 during a discharging process [6].

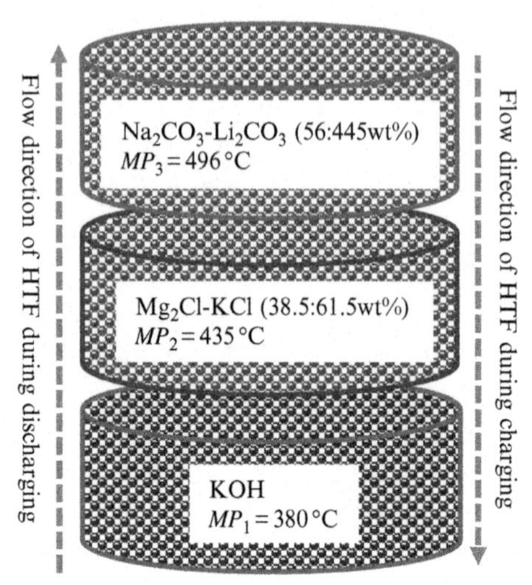

Fig. 5.6 Multiple zones of packed PCM capsules in different melting point and properties [5].

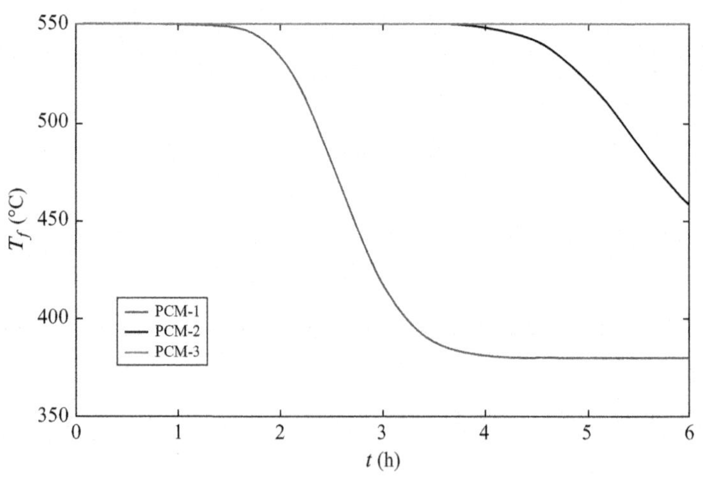

Fig. 5.7 Schematic illustration of the fluid temperature taken at the top of each PCM zone during a discharging process for the PCM packed bed shown in Fig. 5.6 [6].

REFERENCES

[1] Van Lew JT, Li P, Chan CL, Karaki W, Stephens J. Analysis of heat storage and delivery of a thermocline tank having solid filler material. ASME J Sol Energy Eng 2011;133:021003.

[2] Li P, Van Lew J, Karaki W, Chan C, Stephens J, Wang Q. Generalized charts of energy storage effectiveness for thermocline heat storage tank design and calibration. Sol Energy 2011;83:2130–43.

[3] Xu B, Li P, Chan C. Energy storage start-up strategies for concentrated solar power plants with a dual-media thermal storage system. ASME J Sol Energy Eng 2015;137:051002 [12 pages].
[4] Xu B, Li P, Chan C, Tumilowicz E. General volume sizing strategy for thermal storage system using phase change material for concentrated solar thermal power plant. Appl Energy 2015;140:256–68.
[5] Dinter F, Geyer MA, Tamme R. Thermal energy storage for commercial applications: a feasibility study on economic storage systems. Berlin Heidelberg: Springer-Verlag; 1991.
[6] Xu Ben, Zhao Yawen, Chirino Hermes, Li Peiwen. Parametric study of cascade latent heat thermal storage system for concentrating solar power plants In: Submitted to proceedings of the ASME 2017 power and energy conference and exhibition, power energy 2017-3096, June 26–30, Charlotte, NC; 2017.

CHAPTER 6

Volume Sizing for Desired Energy Storage Tasks

Contents

Abstract

This chapter presents methods for calculating the volume of the thermal storage containers/tank to meet the demand of thermal storage for various approaches to thermal storage, including: heat transfer fluid (HTF) thermal storage, solid and HTF dual-media sensible thermal storage, phase change material (PCM) and HTF dual-media latent heat thermal storage, as well as sensible and PCM (latent heat) combined media with HTF thermal storage. Whereas the thermal storage using only HTF is a single medium approach that has ideal energy storage efficiency, other approaches of dual-media thermal storage involve heat transfer between the HTF and the other medium and thus have energy storage efficiency less than 100%. Analysis and computation of the heat transfer between the HTF and the other medium are needed in order to size the volume of the thermal storage container.

Keywords: Determining mass flow rate for thermal storage, Heat transfer fluid, Solar thermal power plant, Thermal storage volume determination, Thermal storage media

Thermal Energy Storage Analyses and Designs
http://dx.doi.org/10.1016/B978-0-12-805344-7.00006-7

6.1 REQUIRED MASS FLOW RATE OF HEAT TRANSFER FLUID FOR POWER DEMAND

A thermal storage system essentially provides a hot heat transfer fluid (HTF) and releases the heat to a destination. In this heat releasing process, the HTF experiences a change of temperature from a high temperature T_H to a low temperature T_L. Vice versa, when a heat storage process is concerned, the HTF experiences a change of its temperature from a low temperature T_L to a high temperature T_H. The mass flow rate of the HTF must satisfy the following requirement if the demand of thermal energy, $\dot{Q}_T$, is known:

$$\dot{Q}_T = \dot{m} \cdot C_f(T_H - T_L) \tag{6.1}$$

If a concentrated solar thermal power plant outputs an electrical power of $\dot{Q}_P$, the needed thermal energy can be determined as:

$$\frac{\dot{Q}_P}{\eta_{th}} = \dot{Q}_T = \dot{m} \cdot C_f(T_H - T_L) \tag{6.2}$$

where $\dot{Q}_T$ is the required thermal energy rate that the HTF needs to provide; η_{th} is thermal efficiency in the thermal power plant that relies on the thermal energy from the HTF; $\dot{Q}_P$ is output electrical power from the concentrated solar thermal power plant. From Eq. (6.2), the mass flow rate of the HTF can be decided as follows:

$$\dot{m} = \frac{\dot{Q}_P}{\eta_{th} C_f(T_H - T_L)} \tag{6.3}$$

It is important to note that no matter what thermal storage approach is applied, the mass flow rate of the HTF must satisfy Eq. (6.3) in order to deliver the desired electrical power in a power plant relying on stored thermal energy. Therefore, the mass flow rate of the HTF is independently decided, regardless of the thermal storage approach.

6.2 DETERMINING THE THERMAL STORAGE VOLUMES FOR DESIRED QUANTITY OF ENERGY STORAGE

The analysis and estimation of the thermal storage volume needs to consider various approaches of thermal storage. The approaches include using an HTF alone for thermal storage, using solid and HTF dual media for sensible thermal storage, using a phase change material (PCM) and HTF dual media for latent heat thermal storage, and using sensible and PCM (latent heat) combined media and HTF for thermal storage.

The analysis begins with the case of thermal storage using only an HTF, which is a single medium approach with ideal energy storage efficiency.

6.2.1 HTF as the Only Thermal Storage Medium

If the HTF is used as the only energy storage medium, the needed mass flow rate is still equal to $\dot{m}$ from Eq. (6.3). The total mass and the corresponding volume of the HTF can be decided using:

$$M_{ideal} = \Delta t \cdot \dot{m};\; V_{ideal} = \Delta t \cdot \dot{m}/\rho_f \tag{6.4}$$

where $\dot{m}$ is decided based on Eq. (6.3). It has been previously discussed that when the HTF is used as the sole thermal storage medium, the energy storage efficiency can reach 100% in an ideal situation, assuming that the storage tank is well thermally insulated.

6.2.2 Solid and HTF Dual-Media Sensible Thermal Storage

When solid material forms a packed bed and an HTF flows through it, the needed mass flow rate will still equal $\dot{m}$ as obtained from Eq. (6.3). The thermal storage volume should satisfy the following equation [1]:

$$\left\{\left[(\rho C)_s \cdot (1-\varepsilon) + (\rho C)_f \cdot \varepsilon\right] \cdot V_{min}\right\} \geq \left[(\rho C)_f \cdot V_{ideal}\right] \tag{6.5}$$

where the left-hand side is the total heat capacity of the packed bed (porous media) with HTF filled in the void, and the right-hand side is the ideal case that HTF is the only thermal storage medium and the volume of the storage tank is V_{ideal}. Depending on the density and heat capacity of the solid material relative to those of the HTF, the thermal storage volume V_{min} can be smaller or larger than V_{ideal}.

In order to decide the volume (V_{min}) of the thermal storage tank, a study of the heat transfer between the solid thermal storage material and the HTF has to be conducted. This will predict the HTF temperature discharged from the tank in the thermal discharging process. It is typically required that the discharged HTF temperature be equal to T_H, or no less than a predetermined minimum temperature.

However, because the dual-media thermal storage is not ideal, typically only some percentage of the discharged thermal energy can be withdrawn. This means that if the HTF at a temperature of T_H is charged into a thermal storage tank for a period of time, Δt, the temperature of the HTF discharged from the tank cannot be maintained always at T_H, but drops over the period of time Δt.

Therefore, besides the conditions given in Eq. (6.5), either one of the following two approaches must be applied in order to finalize the proper thermal storage volume and cause the discharged HTF to have the same temperature in the same period of time as for the charging process [2]. These approaches include: (1) charge energy with longer period of time than the required discharging time, (2) charge energy with higher flow rate and discharge at the required flow rate in the same period of charging time. This second approach is usually not recommended, because heat collectors are typically designed based on a given or fixed flow rate.

Another issue during the design analysis of the thermal storage volume is to find the temperature distribution in the thermal storage tank at charged status and discharged status for cyclic operations. For example, the thermal storage system may start from a fully cold state, and after a number of cycles of charging and discharging the temperature distribution after discharging will be different from the initial cold state. This temperature distribution in the thermal storage tank after a discharging is needed as the initial condition for the charging process, and vice versa [3].

To decide the dimensions of a thermocline storage tank, the required operational conditions from the power plant include: the electrical power, the thermal efficiency of the power plant, the desired period of operation based on stored thermal energy, the required high temperature of heat transfer fluid from the storage tank, the low temperature of fluid returned from the power plant, the properties of the heat transfer fluid and the thermal storage material, including the nominal radius of packed pebbles if applicable, as well as the packing porosity in a thermocline tank.

The computational analysis will include the following steps:

(1) Decide on an ideal volume using Eq. (6.4) for the ideal case where only an HTF is used for thermal storage. Once the volume of the ideal thermocline tank is determined, a minimum storage tank volume should be determined using the criterion given in Eq. (6.5). This gives the lower limit of the volume, below which one cannot achieve the desired thermal energy discharge. From the determined tank volume, one can choose diameter, *R*, and the corresponding height, *H*, which will be used in the first trial for energy storage effectiveness analysis.

(2) Use dimensions for the determined minimum volume of the container to conduct one-dimensional heat transfer analysis for one charging and one discharging, and examine the discharged fluid temperature. This will make sure that the fluid temperature is above the minimum required temperature in the desired period of discharging. Typically, the minimum volume cannot meet the requirement for the system to have a discharged fluid temperature the same as that of the charged fluid. By considering an enlarged tank volume and also a longer charging time (compared to the discharging time), one may achieve the goal that the discharged fluid temperature from the thermal storage system is the same as that of the charged fluid over the desired period of time.

(3) Since a thermal storage system needs to operate over many charging and discharging cycles, it is best that the system design analysis take the case of steady periodical operation into consideration. During the initial cycle of energy charging and discharging, the storage tank is fully cold and therefore much longer charging is needed. Since the storage system will have residual energy after each discharging, gradually the thermal storage system will warm up until a steady cyclic operational condition is built up. Therefore, the computation needs to consider up to 10 times of charging and discharging [3] in order to reach steady cyclic operation.

These steps for thermal storage volume design are explained in the flow chart shown in Fig. 6.1 [4].

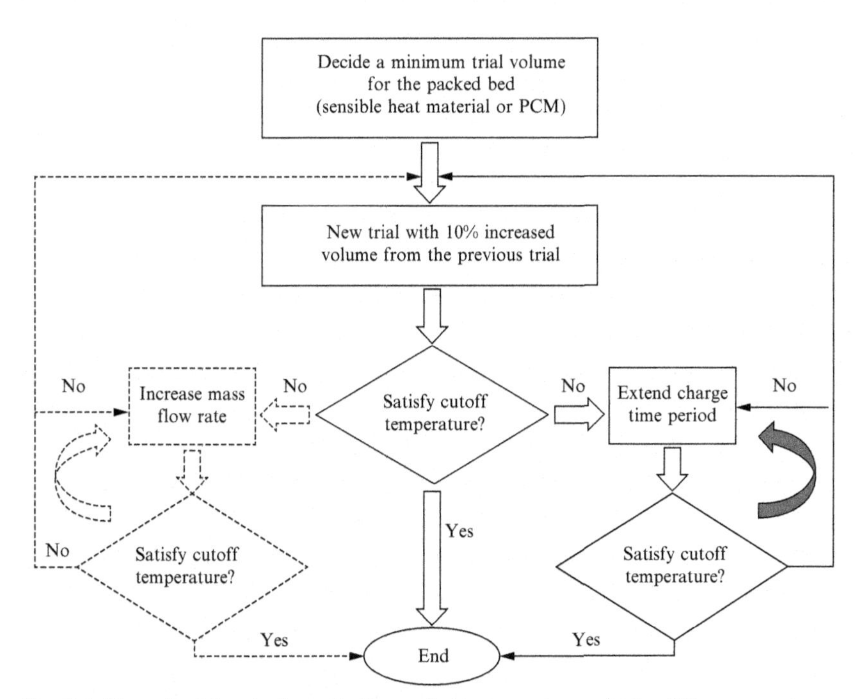

Fig. 6.1 Flow chart for dual-media thermal storage volume design [4].

6.2.3 PCM and HTF Dual-Media Thermal Storage

When an encapsulated PCM is packed in a tank and an HTF flows through it, the needed mass flow rate will still equal $\dot{m}$ as obtained from Eq. (6.3). Because of the latent heat of the PCM, the thermal storage volume should satisfy the following equation:

$$\left\{\rho_{PCM}\Delta H_{fusion}(1-\varepsilon)+(\rho C)_f(T_H-T_L)\varepsilon\right\}\cdot V_{PCM-min} \geq (\rho C)_f\cdot V_{ideal}(T_H-T_L) \tag{6.6}$$

where ΔH_{fusion} is the latent heat of the PCM with the melting point close to or lower than T_H. Assuming that the heat transfer between the PCM and the HTF is ideal, the tank volume, $V_{PCM-min}$, obtained from Eq. (6.6) can be the minimum volume that the heat transfer analysis should start with. Choosing a proper diameter of the storage tank, the height of the tank can be determined based on $V_{PCM-min}$, and heat transfer and energy storage analysis can be conducted using the flow rate determined by Eq. (6.3). The analysis simulates the heat charging and discharging cycles and examines the HTF temperature during a heat discharging process. Typically, if the discharged HTF temperature cannot satisfy the requirement for a certain time period, the $V_{PCM-min}$ will need to be enlarged for the second trial of analysis. In this case, it is also important that an energy charging over a longer period of time than the required discharging time is chosen for the operation of the system.

6.3 EXAMPLES OF SIZING OF THERMAL STORAGE SYSTEM APPLIED TO CONCENTRATED SOLAR THERMAL POWER SYSTEMS

6.3.1 Calibration Analysis for Existing Storage Volume to Meet a Demand

If the dimensions of a storage tank and the operational conditions of a thermal storage system, including the mass flow rate and the high and low HTF temperatures (T_H, T_L) are given, a designer may be required to find a proper time period of energy charge that can satisfy the needed operation time of the thermal energy delivery. The known parameters will be the ideal tank volume, τ_r, as well as H_{CR} at a required operational period of Π_d, as defined in Chapter 4 (Section 4.2).

The first step of the calibration should be the examination of the criterion given in Eq. (6.3), from which a minimum tank volume can be decided. If the minimum tank volume is satisfied, the second step of the calibration is to find a proper ratio of charging time versus discharging time Π_c/Π_d that will allow the energy delivery effectiveness to approach 1.0. The calculated energy storage effectiveness (Eq. (4.16) in Chapter 4) vs. Π_c/Π_d at the given H_{CR} (Eq. (4.12) in Chapter 4) can be quickly found. If at the required Π_d the energy storage effectiveness cannot approach 1.0 at any value of Π_c/Π_d, a new Π_d will have to be selected.

6.3.2 Examples of Storage Tank Size Design for Sensible Energy Storage

Example 1 [1,2]: a pilot solar thermal power plant has 1.0 MW electrical power output at a thermal efficiency of 20%. The heat transfer fluid used in the solar field is Therminol VP-1. The power plant requires high and low HTF temperatures of 390°C and 310°C, respectively. River rocks are used as the filler material and the void fraction of packed rocks in the tank is 0.33. The required time period of energy discharge is 4 h, the storage tank diameter is 8 m. The rock diameter is 4 cm. This example shows how a designer can, upon finding τ_r, Π_d, and H_{CR}, determine whether the design will deliver a high energy effectiveness (Eq. (4.16) in Chapter 4) and how to modify the design. The steps of the analysis are as follows:

First, find a necessary mass flow rate of 25.34 kg/m^3 and an ideal tank height of 9.59 m. The minimum volume of the storage tank can be determined from Eq. (6.5). The ideal volume is used in the first design trial. Using the equations listed in Table 4.1 for spherical rocks, we find the corrected heat transfer coefficient [5,6] to be 32.05 W/m^{2}K. With this information, the values of H_{CR}, Π_d, and τ_r (with definitions in Chapter 4 (Section 4.2)) are obtained as 0.451, 3.03, and 0.0227, respectively. For the condition that $H_{CR} = 0.45$, there is no time ratio Π_c/Π_d that allows the energy delivery effectiveness to approach 1.0. Therefore, the ideal volume will not satisfy the energy storage demand.

One option to improve the ability to store and deliver more energy is to increase the height of the storage tank. When the height is increased to 12 m, the values of Π_d and τ_r change to 2.42 and 0.0181, respectively. It is found that at $H_{CR} = 0.45$ and at the ratio, $\Pi_c/\Pi_d = 1.2$, the energy delivery effectiveness approaches 0.99. This becomes an acceptable design.

Compared to an ideal thermal storage tank, the rock-packed bed thermocline tank uses about 40.0% of the HTF. The longer energy charging time than discharging time indicates that 20% of the charged energy cannot be used, due to the temperature degradation in the discharging process.

Example 2 [1,2]: For the same solar thermal power plant and operational conditions as in Example 1, the thermal storage primary material is molten salt with properties of $\rho = 1680\mathrm{kg/m^3}$, $C_s = 1560\mathrm{J/(kg \cdot K)}$, and $k_s = 0.61\mathrm{W/(m \cdot K)}$. The heat transfer fluid HITEC flows in multiple heat transfer tubes as shown in Fig. 4.3D in Chapter 4. The required time period of energy discharge is 4 h, and the storage tank diameter is 8 m. The study will find the storage tank height.

Following the same procedure as in Example 1, we have a mass flow rate of 40.35 kg/m^3 and the ideal tank diameter and height of 8 m and 6.44 m, respectively. Assume that we have 8448 steel pipes (with an inner diameter of 0.025 m) in the storage tank, and the HTF flows in all the pipes with an equal flow rate. The void fraction ε is found to be 0.33, and the heat transfer surface area per unit of length of the tank is found to be $S_s = 1327\mathrm{m}$. Using the equations listed in Table 4.1 for the case of Tubes, we can find the corrected heat transfer coefficient. The minimum volume of the tank is used in the first trial of the design.

The dimensionless values of H_{CR}, Π_d, and τ_r are 0.522, 3.032, and 0.221, respectively. At H_{CR} of 0.50 (close to 0.522) it is impossible to get an energy delivery effectiveness of 1.0 at all the trialed time ratios of Π_c/Π_d.

To increase the energy delivery effectiveness, a new tank height of 2.1 times the ideal volume is used. This makes the values of Π_d and τ_r 1.444 and 0.105, respectively. Now it is seen that at H_{CR} of 0.50 (close to 0.522) and a time period ratio Π_c/Π_d of 1.2, the energy delivery effectiveness can reach 0.96.

Note that after the tank height is increased, the HTF volume takes 69% of the ideal heat transfer fluid volume. The energy delivery effectiveness is able to reach as high as 0.96. In order to improve the energy delivery efficiency, the heat transfer between fluid and thermal storage material must be improved, for example, using pipes with fins.

Example 2 also indicates that when the heat transfer performance (the multiplication of heat transfer coefficient and heat transfer area) between HTF and thermal storage material is poor, the energy delivery effectiveness can be low. If the temperature degradation of the discharged fluid is a big concern, for example in a power plant, heat transfer enhancements in the thermal storage must be made.

6.3.3 Examples of Storage Tank Size Design for PCM Energy Storage

A parabolic trough CSP plant with 60 MW electrical power output at the thermal efficiency of 35% is taken as an example, based on the systematic design in the work of Biencinto et al. [7]. The HTF used in the solar field is Therminol VP-1. The power plant requires high and low fluid temperatures of 390°C and 310°C, respectively. Depending on the packing scheme, the void fraction ε in a packed bed with spheres of a fixed diameter ranges from 0.26 to 0.476 [8]. In this paper, a void fraction of 0.3 was chosen for all the calculations. The diameter of the encapsulated PCM is 4 cm, and the radius of the storage tank is set as 5 m. The required minimum time period of energy discharge is 6 h. The thermal storage system is assumed to be operated within 100 days or 100 charge/discharge cycles without maintenance. The minimum cutoff temperature is set to be 360°C in a discharge process, below which the HTF cannot be effectively used for power generation. This cutoff temperature is based on the conclusion from Modi and Pérez-Segarra [9] that 30°C below the high temperature is acceptable. All the operating conditions and the properties of the HTF and filler material are listed in Tables 6.1 and 6.2. The details of numerical modeling and computational procedures are referred to Chapter 4 as well as to reference [10].

Based on the properties, efficiency of the power plant, and need for power, one can calculate a necessary total mass flow rate for the thermal storage as follows:

Table 6.1 Operational parameters of a 60 MW CSP plant for 6 h thermal storage [4]

$P_{ele}=60$ MW—Electrical output	$\xi=35\%$—Power plant thermal efficiency
$R=5$m—Radius of storage tank	$T_{cutoff}=360°$C—Cutoff temperature
$T_H=390°$C—High temperature	$T_L=310°$C—Low temperature
$N_{cycle}=100$—Number of cycles	$\Delta t_{discharge}=6$h—Time period of discharge
$\varepsilon=0.3$—Void fraction	$d_r=0.04$m—Diameter of filler material

Table 6.2 Properties of HTF and filler materials [4]

HTF (Therminol VP-1)	
$\rho_f=761\,\text{kg/m}^3$	$C_f=2454\,\text{J/(kgK)}$
$k_f=0.086\,\text{W/(m K)}$	$\nu_f=2.33\times10^{-7}\,\text{m}^2/\text{s}$
PCM-1 (KOH, potassium hydroxide)	
$\rho_r=2044\,\text{kg/m}^3$	$C_{r_s}=1470\,\text{J/(kgK)}$solid
$k_r=0.5\,\text{W/(mK)}$	$C_{r_l}=1340\,\text{J/(kgK)}$liquid
$L=149.7\,\text{kJ}$	$T_{melt}=380°$C

$$\frac{P_{ele}}{\xi} = \dot{m}_{total} C_f (T_H - T_L) \tag{6.7}$$

The required total mass flow rate for a 60 MW electrical supply is calculated as $\dot{m}_{total} = 873.21$ kg/s, which can be divided into four substreams for four tanks, and the mass flow rate $\dot{m}$ in each substream is thus equal to 218.3 kg/s. In the following calculations, only one substream is studied, since the results for one substream will be identical to the other three substreams.

Next, the first step in the design process is to choose an initial trial thermal storage tank volume using Eq. (6.6). The height of the storage tank can be determined as shown in Table 6.3.

The charge time period initially is set to be equal to the discharge time period (6 h). The modeling computation for the thermal performance of the storage system can be obtained. In the next section, the results are examined to obtain an actual storage tank volume, following the general volume sizing strategy.

As a first step, the temperature of outflow HTF versus discharge time at cyclic periodic steady state has been plotted in Fig. 6.2, in which the dashed line represents the cutoff temperature.

It can be observed clearly from Fig. 6.2 that the storage system can only supply HTF above 360°C for approximate 3.5 h, which is much less than the required 6 h. Therefore, based on the general volume sizing strategy, the volume of the trial storage tank is increased by 10%. Also the heat charging time is extended from 6 to 8 h, and a comparison is made in Fig. 6.3.

It is seen from Fig. 6.3 that even if the charging time is extended to 8 h, the 10% increase of the tank height from the trial tank height still would not fully meet the cutoff temperature of above 360°C in 6 h. The 7-h charge with 10% increase of volume can discharge HTF for about 3.9 h with the temperature above 360°C, while the longest time period of 4.1 h is offered by an 8-h charge. Even though these results are better than the 6-h charge at the first-trial tank volume, it is still far from the target of an entire 6-h supply of fluid at temperatures above 360°C. As a result, the next step is to go back and increase the tank volume of the first trial another 10% or more.

Table 6.3 Height of trial storage tank for each PCM when the cutoff temperature is 360°C [4]

PCMs	Tank height of initial trial
PCM-1 (KOH)	25.9 m (total of multiple tanks)

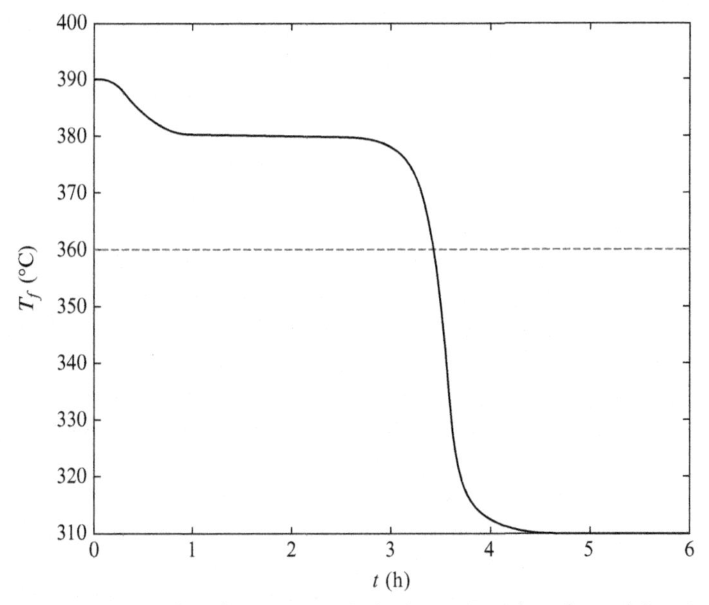

Fig. 6.2 Temperatures of outflow HTF in 6 h discharge based on the tank height of the initial trial analysis [4]

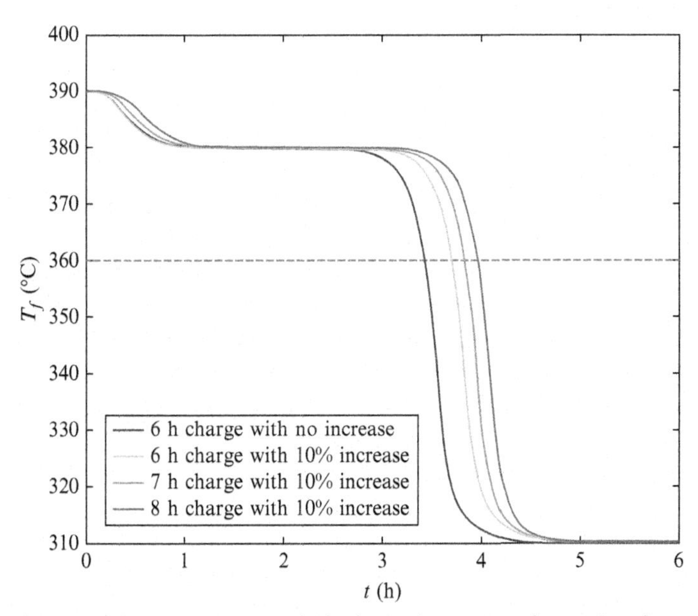

Fig. 6.3 Output HTF temperature in 6 h discharge by varying charge time from 6 to 8 h with 10% increase of the height from the volume of the first trial [4].

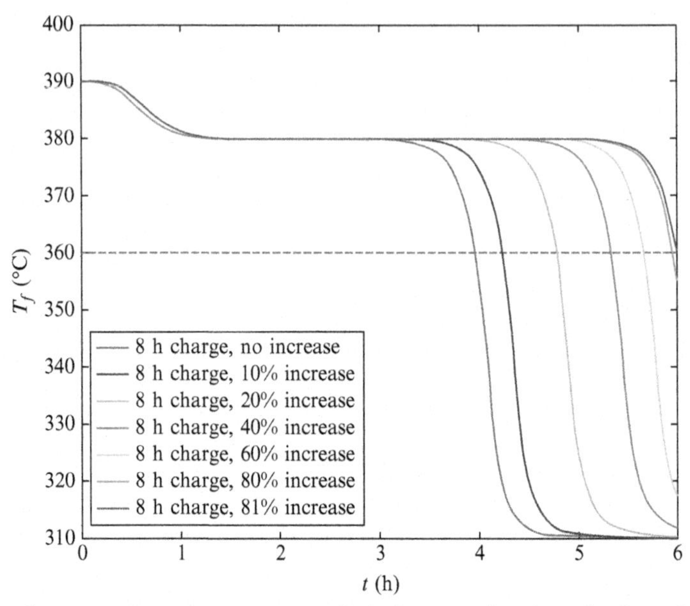

Fig. 6.4 The output HTF temperature in 6 h discharge with enlarged volume from the first trial and 8 h charging time [4].

Following that, the charging time is now chosen as 8 h and the volume is also increased. The results are shown in Fig. 6.4.

Fig. 6.4 demonstrates the output HTF temperature within a 6-h discharge, based on enlarged volume from the first trial storage volume and a fixed charging time of 8 h. It is seen clearly from this figure that before the volume is increased 80%, the output HTF temperature at the end of 6 h of discharge is far from the cutoff temperature of 360°C. Given an 81% increase of the tank volume and 8-h charging time, the results of the discharged fluid temperature could completely satisfy the requirement of above 360°C during the entire discharging time period of 6 h. For this result, the storage tank volume is 3683.5 m^3, which may require multiple tanks in a total height of 46.9 m, if the radius of each tank is fixed at 5 m. In this case, the energy storage efficiency is 88.19%.

For the same thermal storage requirement and PCM material, if the solar radiation collection time cannot be longer than 6 h, the only option to meet the requirement is to further enlarge the storage tank volume. Fig. 6.5 shows the comparison of output HTF temperature during a 6-h discharge with various storage tank volumes at a fixed 6 h of charge. One can observe from the figure that only when a 90% increase of the volume is employed can the

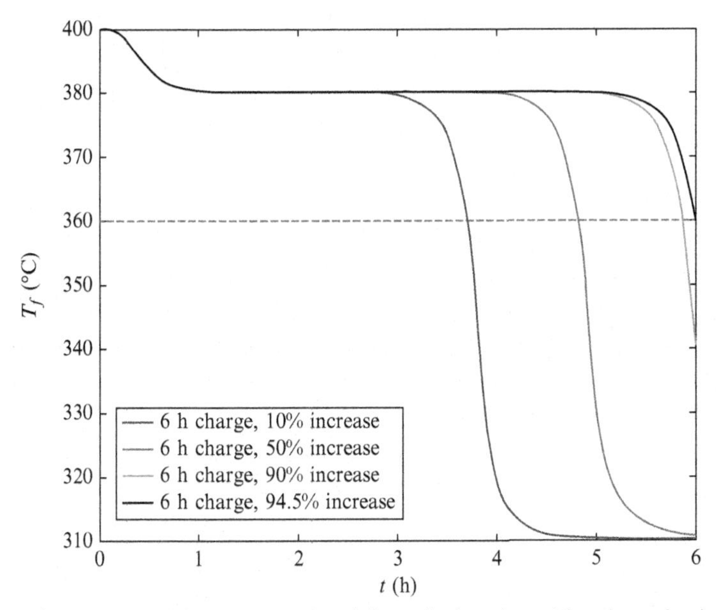

Fig. 6.5 The output HTF temperature in 6 h heat discharging with enlarged volume of storage tank and 6 h heat charging.

output HTF temperature at the end of a 6-h discharge get close to 360°C. At the end of the computation, a 94.5% increase of the trial storage tank volume was found to meet the requirement that the discharged fluid temperature be above 360°C during the entire discharge time period of 6 h. For this result, the tank volume is 3954.2 m^3, and the total height of multiple tanks is 50.4 m if the storage tank radius is fixed at 5 m. The energy storage efficiency for this case is 87.61%. This efficiency is close to that of the last case, which has energy storage efficiency of 88.19%. This means that with a larger volume of the storage tank, one can use a shorter charge time of 6 h to achieve the same goal.

These two cases show that the general volume sizing strategy offers designers multiple options to decide on an appropriate storage tank volume and operation scheme to maintain the HTF temperature above the cutoff temperature during a required 6 h of discharge.

6.4 COLD STORAGE ANALYSIS FOR DESIRED COOLING

Due to development of PCM encapsulation technology, a low melting PCM has been widely used for cold storage in recent years [11]. This is

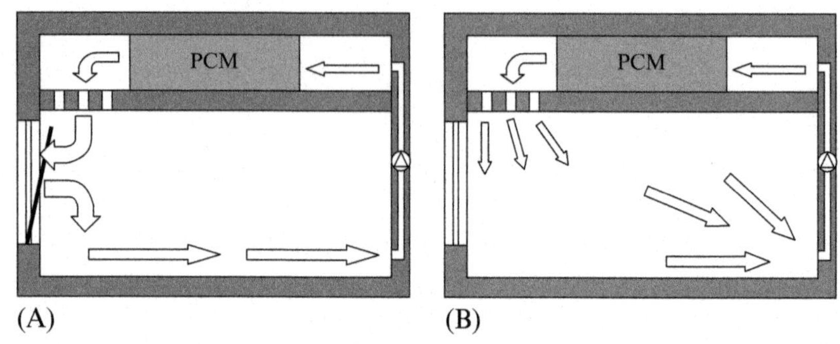

Fig. 6.6 Using PCM for air-conditioning. (A) Cooling of PCM at night; (B) cooling of room in the day. *(Courtesy of Butala V, Stritih U. Experimental investigation of PCM cold storage. Energy Build 2009;41:354–59).*

because encapsulation of the PCM is very important in creating a large heat transfer surface area.

Cold storage can be applied for air conditioning [12], food and vegetable cold storage, transportation, etc., as seen in Fig. 6.6. The heat transfer and energy storage analysis for cold storage is the same as that for heat thermal storage. In this case, a heat charging process is for cold discharging, and the heat discharging process is for the cold charging process. Readers also need to note that the coordinates of all the one-dimensional models presented in this book follow the flow direction of the HTF.

Depending on the application of cold storage, one may also set a limit on the HTF temperature, above which the cold discharging process has to stop.

REFERENCES

[1] Li P-W, Van Lew J, Chan C-L, Karaki W, Stephens J, O'Brien JE. Similarity and generalized analysis of efficiencies of thermal energy storage systems. Renew Energy 2012;39:388–402.
[2] Li P, Van Lew J, Karaki W, Chan C, Stephens J, Wang Q. Generalized charts of energy storage effectiveness for thermocline heat storage tank design and calibration. Sol Energy 2011;83:2130–43.
[3] Van Lew JT, Li PW, Chan CL, Karaki W, Stephens J. Analysis of heat storage and delivery of a thermocline tank having solid filler material. J Sol Energy Eng 2011;133:021003.
[4] Xu B, Li P, Chan C, Tumilowicz E. General volume sizing strategy for thermal storage system using phase change material for concentrated solar thermal power plant. Appl Energy 2015;140:256–68.
[5] Xu B, Li P-W, Chan CL. Extending the validity of lumped capacitance method for large Biot number in thermal storage application. Sol Energy 2012;86(6):1709–24.

[6] Li P, Xu B, Han J, Yang Y. Verification of a model of thermal storage incorporated with an extended lumped capacitance method for various solid-fluid structural combinations. Sol Energy 2014;105:71–81.
[7] Biencinto M, Bayón R, Rojas E, González L. Simulation and assessment of operation strategies for solar thermal power plants with a thermocline storage tank. Sol Energy 2014;103:456–72.
[8] Conway JH, Sloane NJH. Sphere packings, lattices and groups. 3rd ed. New York: Springer-Verlag; 1998. 0-387-98585-9.
[9] Modi A, Pérez-Segarra CD. Thermocline thermal storage systems for concentrated solar power plants: One-dimensional numerical model and comparative analysis. Sol Energy 2014;100:84–93.
[10] Tumilowicz E, Cho Lik C, Li P, Xu B. An enthalpy formulation for thermocline with encapsulated PCM thermal storage and benchmark solution using the method of characteristics. Int J Heat Mass Transf 2014;79:362–77.
[11] Veerakumar C, Sreekumar A. Phase change material based cold thermal energy storage: materials, techniques and applications—a review. Int J Refrig 2016;67:271–89.
[12] Butala Vincenc, Stritih Uros. Experimental investigation of PCM cold storage. Energ Buildings 2009;41:354–9.

CHAPTER 7

Thermal Storage System Construction and Mechanical Issues

Contents

Abstract

Small-scale and low-temperature thermal storage systems have fewer concerns with structural designs and construction. However, for large-scale and high-temperature thermal storage systems such as those used in concentrated solar power plants, the structural design and stress analysis for the tanks, tank foundations, connections of pumps with motors through the tank wall, and thermal expansion are critical issues that have to be carefully considered. Due to the high temperature and large quantity of high temperature heat transfer fluid in a container, safety and reliability of containers, pipes, pumps, and other facilities is very critical. This chapter briefly discusses solutions to these issues.

Keywords: Heat transfer analysis, Thermal storage container, Thermal storage materials, Thermal storage system construction, High temperature thermal storage

7.1 PUMP AND MOTOR CONNECTION

There are different ways to set the motor and pump connection for a high temperature thermal storage flow loop. While a pump submerges in the high temperature fluid, the electric motor for the pump must avoid overheating. This requires that a sufficiently long connection between the pump and the motor be used.

If the motor sits on top of the thermal storage tank and the pump is submerged in fluid at the lower end of the tank [1], the connection of motor and

Thermal Energy Storage Analyses and Designs
http://dx.doi.org/10.1016/B978-0-12-805344-7.00007-9

pump through the tank roof is subject to relatively low pressure, which is helpful with regard to the sealing issue. The disadvantage of this setting (may be called as vertical pump setting) is the long connection between the motor and pump. From a construction point of view, putting the motor on top of the tank may cost more.

If a pump and motor are both set at the lower end of a storage tank, a very good seal under high temperature for the shaft connection between the pump and the motor is needed to avoid leakage of the heat transfer fluid (HTF). Sealing technologies such as those used for high temperature gas turbines may be applicable. Locating the pump and motor at ground level is the advantage of this setting.

7.2 HIGH PRESSURE AT BOTTOM FOR THERMAL STORAGE TANKS

Typically thermal storage materials are expected to have high density compared to water [2–4], which makes the pressure at the bottom of a storage tank rather high and the tank bottom and foundation must be able to withstand the compression strength. Due to the need to resist a significant pressure, the welding of the tank bottom to the side wall is critically important to the safety of the thermal storage tank. Onsite detection of weld defects is typically needed in construction of thermal storage tanks.

The pump head needed to feed the HTF to the power tower is also significant, due to the high density of the fluid, even though most HTFs do not have high vapor pressure.

7.3 THERMAL EXPANSION AND THERMAL CYCLES

A thermal storage tank may experience significant thermal expansion due to the large change of temperatures from a cold state to an operational state and also during the cyclic thermal charging and discharging processes. Mismatch of the thermal expansion of the metallic tank material and the thermal protection material is also an important factor that needs to be considered when designing the system.

7.4 FREEZING AND MELTING OF FLUID IN TANKS AND PIPES

In a solar thermal power plant, some worst-case weather scenarios may occur so that the thermal storage tank may have excessive loss of heat and the thermal storage material may freeze. To avoid freezing of thermal storage material, an important practice is to start from the design stage. One needs to predict (calculate) the heat loss under the worst possible weather conditions and check whether a large part of the tank could be frozen. The thermal insulation should be correspondingly designed to reduce heat loss. Based on the heat loss analysis, an approach for thermal insulation or reheating of the fluid could be taken.

It is very important that the heat transfer analysis and prediction of the temperature field inside the thermal storage tank help the design for thermal protection of thermal storage containers. Commercial tools for numerical analysis of transient heat transfer are widely available. Therefore, computation and simulation can be conducted to find out the transient processes of heat loss, temperature change, and the possibility of the fluid in a tank freezing.

7.5 LOW CORROSIVITY AND TOXICITY OF THERMAL STORAGE MATERIALS

Although the thermal storage material and HTF are cautiously protected in containers, any leakage and spill-off can still be a concern if the fluid is corrosive and toxic. Therefore, it is always desirable that HTFs have low cost, low toxicity, and low corrosion to containers and pipes. The safety data (toxicity and flammability) of mineral oil heat transfer fluids such as Xceltherm and Therminol VP1 are available in Refs. [5, 6]. These fluids are categorized as low toxicity materials, but they are flammable and safe treatment is important to prevent fire. The toxicity of nitrate molten salts and chloride molten salts heat transfer fluids is also generally low. Safety information for salts such as $NaNO_3$, KNO_3, KCl, $MgCl_2$, are widely available elsewhere, such as databases [7, 8].

REFERENCES

[1] Pacheco JE, Showalter SK, Kolb WJ. Development of a molten salt thermocline thermal storage system for parabolic trough plants. J Solar Energy Eng 2002;124(2):153–9.

[2] Li C, Li P, Wang K, Molina EE. Survey of properties of key single and mixture halide salts for potential application as high temperature heat transfer fluids for concentrated solar thermal power systems. AIMS Energy 2014;2(2):133–57.

[3] Li P, Molina E, Wang K, Xu X, Dehghani G, Kohli A, et al. Thermal and transport properties of NaCl-KCl-$ZnCl_2$ eutectic salts for new generation high-temperature heat-transfer fluids. J Sol Energy Eng Trans ASME 2016;138:054501-1–8.

[4] Yuanyuan L, Xiankun X, Xiaoxin W, Peiwen L, Qing H, Bo X. Survey and evaluation of equations for thermophysical properties of binary/ternary eutectic salts from NaCl,

KCl, $MgCl_2$, $CaCl_2$, $ZnCl_2$ for heat transfer and thermal storage fluids in CSP. Solar Energy 2017; http://dx.doi.org/10.1016/j.solener.2017.03.019.
[5] http://www.radcoind.com/wp-content/uploads/SDS-XCA.pdf.
[6] http://www.americasinternational.com/msds/eastman/Therminol%20VP1.pdf.
[7] https://www.sciencelab.com/msds.php?msdsId=9927232.
[8] http://www.ilo.org/safework/info/publications/WCMS_113134/lang–en/index.htm.

CHAPTER 8

Thermal Insulation of Thermal Storage Containers

Contents

Abstract

The insulation of the storage tank is paramount to ensure the operation of electricity production during the night or periods of no sun. Heat transfer exists whenever there is a temperature difference and can occur in three different modes, conduction, convection, and radiation. The purpose of the insulation is to reduce the heat loss to a minimum. The temperature of the storage tank is on the order of 600 K while the ambient temperature is around 300 K. This very large temperature difference causes a large heat loss. A combination of multiple thermal insulation layers, with each serving different purposes (withstand high temperature or super insulating), is necessary to address this problem. It is a formidable task to insulate a storage tank properly.

Keywords: Insulation materials, Thermal conductivity, Thermal insulation, Thermal storage tank

Throughout its history, thermal insulation has been focused on protecting humans from the elements [1]. Thermal insulation, in particular, creates a thermal barrier to regulate temperature. Once upon a time, humans used natural materials such as animal skins, wool, and straw as insulation for clothing as well as in their habitats. We progressed to more durable materials, such as soil, stone, and wood. For example, adobe was used for many, many years by indigenous people of the Americas. As the building materials shifted to stronger types, e.g., steel and concrete, extra thermal insulation layers became a necessity. Natural materials were exploited and synthetic ones were developed. The exponential increase in energy consumption and its detrimental environmental impacts have caused a wave of innovations in more efficient uses of our resources. In addition, thermal insulation plays an important role in many energy production and manufacturing applications. For instance, firebricks and mineral wool are commonly used in high

Thermal Energy Storage Analyses and Designs
http://dx.doi.org/10.1016/B978-0-12-805344-7.00008-0

temperature applications. The demand for better and better thermal insulation has caused the ongoing development of ultralow synthetic thermal insulations, such as aerogel.

There are many different thermal insulation materials. These materials are characterized by low thermal conductivity, usually below 1 W/m K. It is also true that low thermal conductivity is associated with low density and air gaps because air (or gas) has low thermal conductivity, around 0.02 W/m K [2,3]. A typical thermal insulating material reduces the heat flow by the microscopic dead air cells. However, these voids, if not properly isolated from air infiltration, are prone to convection loss. Furthermore, radiation may play a role, especially when the temperature is high. The lower the thermal conductivity, the better thermal insulation material it is. On the other hand, considerations should be given to environmental and public health aspects, fire hazard, moisture effects, and maintenance schedules.

A review of the traditional, state-of-the-art and future thermal insulation can be found in reference [4] by Jelle. Thermal insulation can be categorized according to inorganic materials and organic materials. Inorganic materials are glass, rock, slag wool, calcium silicate, bonded perlite, vermiculite, and ceramic products. Organic materials include cellulose, cotton, wood, pulp, cane, synthetic fibers, cork, foamed rubber, polystyrene, polyethylene, polyurethane, polyisocynurate and other polymers. They can be either in fibrous form or foam structure. Vacuum insulation panels are an open porous core of fumed silica enveloped by layers of metallized polymer laminate, and the corresponding thermal conductivity is below 0.005 W/mK. Gas-filled panels use a less thermal conductive (than air). Although the thermal conductivity increases, however, the structure of the porous core does not have to withstand the vacuum.

Existing insulation materials that can withstand high temperatures are abundant, such as firebricks, calcium silicate, foamglass, mineral wool, and fiberglass.

Firebrick can withstand high temperatures and usually has a low thermal conductivity. It is widely used in wood-fired kilns and furnaces. Firebrick is a mixture of alumina and silica. By varying the composition, the thermal conductivity can be adjusted. The thermal conductivity is around 1 W/mK and it can operate at high temperatures up to 1500 K.

Foamglass is a cellular glass formed mainly from sand, limestone, and soda ash. This mixture is melted and crushed into powder form for sintering. When a small amount of finely ground carbon-black is added, carbon dioxide is formed and creates insulating bubbles. The thermal conductivity is around 0.05 W/mK and it can operate at temperatures up to 750 K.

Mineral wool is made from molten glass, stone or slag by spinning to form a fiber-like structure. The thermal conductivity is around 0.5 W/mK and it can operate at temperatures up to 1000 K.

Silica aerogel is formed by a process proposed by Schwertfeger and Schmidt [5]; starting with water glass, silation is carried out in the water phase separation of the hydrogel, phase separation of gelwater and solvent. The thermal conductivity is very low around 0.01 W/mK and it can operate at high temperatures up to 773 K.

Using multiple layers of structural and insulating materials is a potential solution to thermal storage tank construction. Gabbrielli and Zamparelli [6] presented an optimal design procedure for internally insulated, carbon steel molten salt thermal storage tanks for parabolic trough solar power plants. They proposed a tank construction using multiple layers of flexible stainless steel liner, firebrick, and ceramic fiber insulating material, while the foundation is constructed of fine sand, insulating firebrick, foamglass, cooled reinforced concrete, poured concrete slab, and foundation piles. They modeled the total heat loss by a thermal resistance model. FEM, SAP2000 software, is used to perform stress analysis. Economic analysis is also performed, to find the optimal design. A viable, inexpensive, and reliable solution was predicted.

To come up with better thermal insulation, engineers and scientists have been using nanotechnology recently [7]. A graphite powder additive is added to expanded polystyrene products and it reduces the thermal conductivity by 20%. Aerogel with its nanostructure is a very good thermal insulator. Nanoceramic thermal insulation coating is also another ongoing effort. The research in thermal insulation will continue. Special insulations targeted for specific applications can be developed. The future is full of possibilities.

REFERENCES

[1] Bozsaky D. The historical development of thermal insulation materials. Period Polytech Archit 2010;42(2):49–56.
[2] Al-Homoud MS. Performance characteristics and practical applications of common building thermal insulation materials. Build Environ 2005;40:353–66.
[3] Papadopoulos AM. State of the art in thermal insulation materials and aims for future developments. Energy Build 2005;37:77–86.
[4] Jelle BP. Traditional, state-of-the-art and future thermal building insulation materials and solutions—properties, requirements and possibilities. Energ Buildings 2011;43:2549–63.

[5] Schwertfeger F, Frank D, Schmidt M. Hydrophobic waterglass based aerogels without solvent exchange or supercritical drying. J Non-Cryst Solids 1998;225:24–9.
[6] Gabbrielli R, Zamparelli C. Optimal design of a molten salt thermal storage tank for parabolic trough solar power plants. J Sol Energy Eng 2009;131:041001.
[7] Bozsaky D. Special thermal insulation methods of building constructions with nanomaterials. Acta Tech Jaur 2016;9(1):29–41.

INDEX

Note: Page numbers followed by *f* indicate figures, and *t* indicate tables.

Printed in the United States
By Bookmasters